~

I. P. Kovalyov

SDMA for Multipath Wireless Channels

Springer
Berlin
Heidelberg
New York
Hong Kong
London
Milan
Paris
Tokyo

Engineering ONLINE LIBRARY

springeronline.com

I. P. Kovalyov

SDMA for Multipath Wireless Channels

Limiting Characteristics and Stochastic Models

 Springer

Igor P. Kovalyov

Nizhny Novgorod
State Technical University
Minina st. 24
603600 Nizhny Novgorod
Russia

E-mail: kovalyov@unc.sci-nnov.ru

ISBN 3-540-40225-X Springer Verlag Berlin Heidelberg New York

Cataloging-in-Publication Data applied for
Bibliographic information published by Die Deutsche Bibliothek
Die Deutsche Bibliothek lists this publication in the Deutsche Nationalbibliografie;
detailed bibliographic data is available in the Internet at <http://dnb.ddb.de>.

Springer-Verlag is a part of Springer Science+Business Media

springeronline.com

© Springer-Verlag Berlin Heidelberg New York 2004
Printed in Germany

Typesetting: Digital data supplied by author
Cover design: design & production GmbH, Heidelberg
Printed on acid-free paper 62/3020hu – 5 4 3 2 1 0 –

Preface

Mobile communication systems are here to stay and are being used more and more. They find extensive application not only for voice but also for data transmission, with data transfer rates steadily growing. It is safe to say with reasonable confidence that communication systems have gone from great height – classical antennas installed on masts and roofs – to portable systems with small-size antennas, which are in wide use at present.

Transmitter-to-receiver wave propagation in mobile communication systems, in distinction to stationary ones, follows a different pattern. The radio waves in such systems take many paths to arrive at the receiver, bouncing off buildings, the earth's surface, pillars and other obstacles. Such a wireless channel is usually referred to as a multipath one. Multipathing is a special feature of mobile communication systems; the direction of wave arrival and wave polarization in such channels cannot be predicted.

Active harnessing of digital devices in communication systems promotes continuous improvement of their quality, performance and efficiency. Development of space diversity systems is speculated to be one of the most promising avenues in improving performance of communication systems. Such are the SDMA (spatial division multiple access) systems, which distinguish the signal sources not only by the time slots (TDMA), and the frequency band involved (FDMA) or by a unique key used (CDMA), but also by the location of the signal sources. The systems based on spatial processing can be spoken of as "stereo" communication systems by analogy with stereo audio systems.

A host of books and journal articles propose and examine a wealth of algorithms including a great many algorithms pertaining to spatial processing. In such circumstances it is vital to determine and understand the theoretical limit of performance improvement. It is of interest to draw the line that cannot be crossed in developing a communication system, however complex and sophisticated the employed algorithms may be. Finding such a limit is one of the objectives of the book. In other words, the challenge was to extend Shannon's channel capacity equation to cover the wireless channels that use spatial signal processing.

In dealing with the limit capacity problem, attention is focused on the fact that the information carrying medium in a wireless channel is the electromagnetic field. For this reason the calculations are based on the amount of information that can be extracted during recording an electromagnetic field at some region of space.

As a matter of convenience, it is assumed that the entirety of information contained in the electromagnetic field can be extracted by a conceptual antenna, which here is referred to as a "potential antenna." The limiting capacity determined with the help of this antenna makes the limiting capacity of a MIMO (multiple input multiple output) communication system with an infinite quantity of inputs and outputs. The inputs of the system are the primary and secondary wave sources that are located at a great distance from the reception area. The system outputs are the pickups for the electromagnetic field located in the receiving area.

The book offers SNR (signal-to-noise ratio) dependence plots for the limit capacity and the optimal number of subchannels of such MIMO systems. An examination of the plots discloses that in a multipath wireless channel the number of subchannels rapidly increases as the size of the receiving area grows larger. The obtained results suggest that new antenna designs need to be developed for compact devices based on the MIMO technology. It is impossible to bring the data transfer rates into proximity with the performance limit using a multielement antenna array made of conventional elements. The performance characteristics of the potential antenna provide a good benchmark that practical implementations must aim at. How successful a particular antenna is can be judged by comparing its operating characteristics against the limiting performance estimations presented in the book.

The majority of the graphs in the book are based on the assumption that the number of propagation paths is immense. In reality their number is quite limited. The book includes several graphs that reveal a decrease of the limiting capacity and the optimal number of subchannels brought about by a limited number of the propagation paths. Such illustrative graphs are few, though, as it is impossible to estimate all practical conditions of interest, and an attempt to do so would result in an overblown book. Instead, in Chap. 2 the reader is provided with a tool allowing him to estimate the situations of interest on his own. In doing so, one can manipulate the number of wave sources, their positioning in space, the size and the shape of the receiving area, as well as other initial data of the problem.

Statistical models of two-dimensional (2-D) and three-dimensional (3-D) wireless channels are another issue under consideration in the book. Statistical modeling is indispensable in the development of communication systems and evaluation of their performance. Furthermore, communication systems based on spatial processing are especially demanding from the standpoint of channel modeling. The model should provide a means of estimating the systems employing 3-D space diversity, polarization diversity, and allow for motion of mobile users. Application of the conventional method, which consists in summing up all the plane wave fields (rays), results in a cumbersome model, especially in the case of the three-dimensional channel.

It is suggested in the book that functions, identified in mathematics and electrodynamics as spherical harmonics, be used for modeling the 3-D wireless channel. A 2-D countertype of the spherical harmonics is used in modeling the 2-D channel. All the components of the electric and magnetic fields are determined by a sum of the fields of the spherical harmonics with random coefficients.

It is safe to assume that the proposed statistical model will prove convenient for description of a multipath field in a limited region of space. It turns out that the number of the random coefficients does not increase with a rise in the number of propagation paths and rapidly diminishes with a shrinking receiving area. This is why with a receiving area of a limited size the modeling time proves to be reasonably short even for numerical evaluation of a random field in 3-D space.

This book is not a topical overview, but rather presents a new approach for describing multipath wireless channels. For this reason, the list of the reference sources is not long, and mainly includes the publications, the results of which are cited in the text. A description of modern MIMO technology and digital radio architecture can be found in M. Martone "Multiantenna Digital Radio Transmission", Artech House, 2002 ; N. Blaunstein and J.B. Anderson "Multipath Phenomena in Cellar Networks", Artech House, 2002. These books also contain an extensive bibliography.

What's in the book

The book includes eight chapters. The first chapter is introductory, and is dedicated to describing the multipath wireless channel and providing a detailed statement of the problems to be tackled.

The second chapter presents without derivation the essential estimation expressions of the book. If the reader's interest is in numerical evaluation and not in the intricacies of derivation he may confine himself to reading the first two chapters only. The sections of Chap. 2 are not linked to each other. To estimate the channel limiting capacity for a variety of the wave source quantities and layouts, it is enough to apply the equations of Sect. 2.1. The shape of the receiving area can be varied during computations too. The reader can use the relationships of Sect. 2.2 for modeling a random multipath field.

Section 2.3 covers the two-dimensional channel, and the 3-D wireless channel is considered in Sect. 2.4. Computations by the equations of Chap. 2 do not imply thorough knowledge of the electrodynamics and spherical harmonics in the reader. Using the presented formulae and a computer the reader may independently perform calculations of interest to him.

Chapter 3 contains information known from other literature and used later in the book. It tackles matrix channels and describes the limit capacity estimation methods for noises of various types.

Chapter 4 is dedicated to the 2-D wireless channel. It discloses derivation of the equations that can be used for estimating the limiting capacity of the channel with receiving areas of diverse shapes. Chapter 4 draws up a statistical model of the 2-D wireless channel and examines its relation to the traditional ray-oriented one. Readers who are interested in 2-D channels only may find it sufficient to read the first four chapters. The material in the first four chapters is simpler than the contents of the remaining four, which are dedicated to an inquiry into the 3-D wireless channel. Apart from the achieved results and deductions the significance of this chapter is in its methodological implications for the subsequent portions of the book.

The transforms illustrated in Chap. 4 by a relatively simple 2-D channel example are later applied to the far more complicated 3-D wireless channel. The book is organized so that its remaining 4 chapters (5 through 8) present the 3-D countertypes of the four sections, 4.1 – 4.4, of Chap. 4.

Chapter 5 provides a body of mathematics useful for examination of 3-D wireless channels. It deals with the basis functions of the three-dimensional channel, i.e., spherical harmonics, and discloses their relation to plane electromagnetic waves.

Chapter 6 contains estimation formulae that can be used to determine the limiting capacity of a 3-D wireless channel.

The results of a numerical evaluation and the SNR dependence plots for the limit capacity are given in Chap. 7.

In Chapter 8 we build a statistical model of the 3-D wireless channel and discuss the statistical properties of the basis function random coefficients. Under consideration are also some modeling results.

Thus, the present book differs from others by use of spherical harmonics for inquiry into multipath wireless channels. Spherical harmonics provide a means of building a new non-ray statistical model of the wireless channel. They also are helpful in estimating the limiting capacity of the multipath channel, in which the conceptual, potential antenna extracts the entirety of the information from the electromagnetic field. It turns out that big spatial dimensions are not a prerequisite for attainment of the limiting capacity in the spatial processing systems. The book shows that it is fundamentally possible to create portable antenna systems with spatial processing and thus achieve data transfer rates that approach the limiting ones.

Acknowledgements

This book is a result of my long time collaboration with MERA Networks. I would like to express my sincere gratitude to Dmitry M. Ponomaryov, the President of MERA Networks, Inc., for his support and advice. I would also like to thank Yuri B. Akimov, the technical writer of MERA Networks, for the translation of the book, my daughter Oksana Smirnova, and my son-in-law Andrew Smirnov for their assistance in writing it.

Please address your comments and questions to the author at E-mail.
kovalyov@meranetworks.com

Igor Kovalyov

Contents

1 Introduction

Chapter 1 presents an overview of multipath wireless channels. The problems of their modeling and estimating the limit capacity are brought forward.

1.1 Multielement Antenna Communication Systems in Multipath Wireless Environment

Wireless channels in mobile communication systems are distinguished by multipath wave propagation. Many monographs have been dedicated to description of multipath channels [1–6].

In a multipath wireless channel the transmitter signal arrives at the receiver via a variety of paths bouncing off buildings, the earth's surface, pillars and other scattering objects. The likelihood of a signal passing through the line-of-sight (LOS) path in such an environment is small. Wave propagation channels from the transmitter to the receiver are illustrated schematically in Fig.1.1.

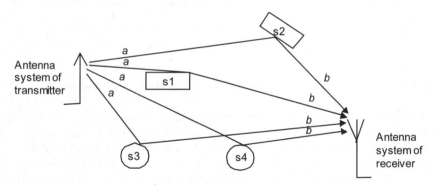

Fig. 1.1. Transmitter-to-receiver wave propagation in a scattering environment (s1, s2,... – scattering objects)

Experimental evidence demonstrates that the number of possible propagation channels can be enormous [7]. Traditionally multipath signal propagation was perceived as a detrimental factor in communication systems and its effect was tenaciously kept as small as possible.

Rather than eliminating the multipath phenomenon the newly established SDMA (space-diversity multiple access) communication systems use it to augment the rate of data transmission. Such systems employ antennas with space-diversity elements. For this reason these systems are sometimes referred to as multielement antenna (MEA) systems [8], space-time coding (STC) systems [9] or Multiple-Input Multiple-Output (MIMO) systems [10]. MEA systems are also referred to as smart antenna communication systems [6 and 11]. The word 'smart' emphasizes continuous assessment of the channel performance taking place in such systems, with changes in the antenna characteristics following the changes in the operational properties of the wireless channel. Smart antennas are capable of distinguishing between signals that match in waveform and coincide in time but arrive from sources located in different spatial points. These may be signals transmitted from various mobile users or distinct elements of the transmission antenna.

In the MEA system the antenna picks up signals from different paths individually. Signals from different paths constitute a diversity of individual spatial subchannels within a wireless channel. Fig. 1.2 provides a diagrammatic sketch of a system with two spatial subchannels.

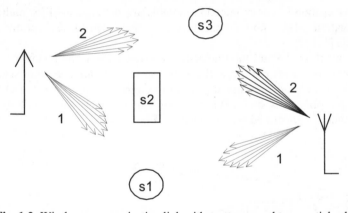

Fig. 1.2. Wireless communication link with scatterers and two spatial subchannels

In this case beams 1 of the transmitting and receiving antenna, traveling via scatterer s1 form one subchannel, while beams 2 form another. The formation of several independent spatial subchannels explains the higher spectral efficiency of MEA communications systems in comparison to communication systems employing single-channel antenna arrays or single-element antennas. The smart MEA systems can ensure discrimination between the space-diverse sources in fluid wireless channels.

The role of the antenna in determining the characteristics of a wireless channel is regarded differently in MEA systems, with a greater emphasis placed upon the antennas' informational properties rather than on their energy qualities, which were the primary subject of analysis before.

The essentials of MEA communication systems were for the first time presented in investigation [12]. Intensive examination and practical implementation of MEA systems commenced in the latter half of the 1990s and was spurred on by the wide use of cellular communication systems. The wealth of theoretical papers that have been published since that time have proposed resolutions to many issues connected with the development of MEA systems. As early as in [13] it was demonstrated that as the number of transmitting and receiving antenna elements increases, so does the specific capacity of the wireless channel which tends to the limit C_∞.

$$C_\infty = \frac{SNR}{\ln 2} \qquad (1.1)$$

In (1.1) C_∞ denotes the system capacity per unit bandwidth (throughput per 1 Hertz of the frequency band), and the signal-to-noise ratio is designated by *SNR*.

The meaning of *SNR* in (1.1) and henceforth throughout this book deserves special explanation. In the MEA system, by applying various excitation techniques to the elements of the transmit antenna and through application of different recombination techniques to the output of the receive antenna elements, it is possible to form a number of single-channel systems with different transmitter-to-receiver signal gains. In such a case *SNR* will also naturally vary from system to system. *SNR* in (1.1) and throughout this book denotes the signal-power-to-noise-power ratio at the output of the single-channel system with the best power gain. In other words, *SNR* indicates the signal-to-noise ratio, which exists when the MEA system provides a perfect single input – single output (SISO) communication link with the entire transmit power entering the wireless channel.

Given *SNR*, the throughput C_1 of the perfect single-channel system (without spatial diversity) is calculated by Shannon's capacity formula [14]

$$C_1 = \log(1 + SNR) \qquad (1.2)$$

Comparison of the SISO system capacity (1.2) to that (1.1) of a system with numerous spatial subchannels is evidence in favor of the latter. With spatial diversity, the capacity is always bigger and this gain in throughput grows with a rise in *SNR*. As a matter of fact, inquiries into MEA systems have demonstrated that the informational properties of the antennas are as important as their energy characteristics.

Investigation of MEA systems shows that as the number of antenna elements increases, so does the capacity of the system, approaching the limit capacity (1.1). However, an increase in the number of antenna elements makes antenna systems grow in size. The question of the maximum capacity that a limited-size antenna can provide is still an open question. The expedience of the attempts to develop portable smart antennas remains a mystery.

In this book we pose and solve the problem of determining the limit capacity of a limited-size antenna.

1.2 Limit Capacity of a Limited-Size Smart Receiving Antenna. Problem Statement

It is of interest to determine the limiting data transfer rate in the multipath wireless channel schematically depicted in Fig. 1.1.

The channel in question may be divided into two legs: the transmitter-to-scatterer leg (*a* in Fig. 1.1) and the scatterer-to-receiver leg (*b*). Since the overall channel throughput does not exceed that of its individual legs, the throughputs of legs *a* and *b* are worth separate consideration.

We shall first concern ourselves with wave propagation from the scatterers to the receiver. This channel is schematically shown in Fig. 1.3.

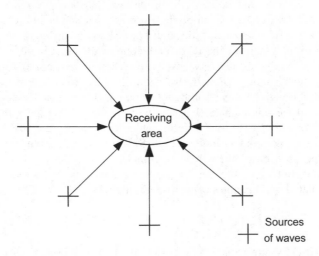

Fig. 1.3. Wave propagation from the scatterers to the receiver

In assessing the limit capacity of the wireless channel we have to abandon the concept of a receiving antenna, substituting a specified limited-size receiving area for it. It is assumed that the receiver extracts the entire amount of information contained in the receive area. The antenna converts the electromagnetic field into signals. The amount of information contained in the electromagnetic field cannot build up during this conversion. Therefore, it is reasonably safe to state that the receiver, which extracts the entirety of information from the input electromagnetic field, ensures the maximum capacity of the wireless channel.

We may assume for convenience that some conceptual antenna, which we will call the potential antenna, extracts the information contained in the electromagnetic field. The potential antenna extracts a maximum of information from the electromagnetic field in the receiving area. Any actual antenna extracts only part of the information. How good a particular antenna is can be judged by the approximation of its operating characteristics to those of the potential antenna.

Therefore, estimating the limit capacity of the wireless channel is one of the problems addressed in this book. The channel's input is represented by the waves (rays) from the scatterers while its output is the image of the electromagnetic field created by the rays in the specified limited area of space.

In the analysis of the limit capacity, we assume that the number of rays (wave sources) is big enough. Formation of each individual ray takes the same amount of power. These assumptions, no doubt, represent an ideal case, whereas in an actual environment the reflections from scatterers differ. It is obvious, though, that presence of weaker scatterers cannot lead to a rise in the information transmission rate. For this reason, in evaluating the maximum information transmission rate, the rays are held to be numerous and of equal power. To put this another way, the wireless channel is assumed to have all the conditions necessary for the maximum capacity.

The analysis of a wireless channel with different conditions presents no special problems, though. Decline in the limit capacity due to decrease in the number of the wave sources is, among other things, analyzed in this book.

Consideration is also being given to narrow-band wireless channels in this book, although the correlation properties of the frequency response of wireless channels will not be discussed. However, the correlation properties of the frequency response are of no importance in estimation of the limit capacity; they become essential only in the development of particular wireless communication systems and in checking their operation algorithms. The limit capacity C_w of a broadband wireless channel may be calculated as the integral of the capacity per unit bandwidth $C(f)$.

$$C_w = \int_{f_{min}}^{f_{max}} C(f)df \qquad (1.3)$$

where f_{min} and f_{max} are the limit frequencies of the broadband wireless channel.

While considering the second constituent of the wireless channel, the transmitter-to-scatterer leg (designated by a in Fig. 1.1) it would be worthwhile to take advantage of duality. That is, we will continue to examine transmission of information from the scatterers to the receive area (instead of concerning ourselves with emission from the transmit area). Then the leg a of the channel will look similar to the leg b, possibly with different parameters. The wave sources (the scatterers) may be situated in some angular domain (Fig. 1.4). Such a radio channel is described as a sector channel in distinction to the wireless channel depicted in Fig. 1.3, which we will designate omnidirectional.

The sector wireless channel provides a lesser means for building spatial subchannels in comparison to the omnidirectional one with the size of the receive area being identical in both the sector and the omnidirectional channel.

However, the size of the receiving area in the sector subchannel is materially bigger. In cellular systems it is represented by the base station with a stationary antenna and not by the mobile user with a portable antenna.

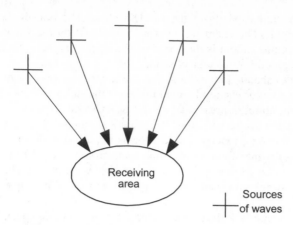

Fig. 1.4. Sector radio channel

Furthermore, current cellular communication systems employ simultaneous operation of several base stations to service distant users. With several base stations operating simultaneously, the sector wireless channel becomes very much like the omnidirectional channel depicted in Fig. 1.3. The capabilities of spatial diversity in case of simultaneous operation of a number of base stations are dramatically augmented and may be exploited for savings in the frequency resource and increase in the number of serviced subscribers.

Let us touch upon the possibility to exploit spatial diversity for users located in close vicinity to the base station when the signal from the transmitter to the receiver follows the LOS path and does not propagate via the scatterers. Creation of spatial subchannels is also possible in an unobstructed environment [15]. The presence of scatterers in the wireless channel is not required. However, in case of an unobstructed environment, spatial separation between the antenna elements is rather big, and the greater the distance between the transmitter and the receiver, the greater the necessary spatial separation. Therefore, if the user is located in close proximity to the base station, spatial separation of the subchannels makes sense. This permits savings in the frequency resource and increase in the number of users.

Noteworthy are multiple reflections in the scattering environment. Though only single reflections are illustrated in Fig. 1.1 when spatial diversity is the case, the rays that arrive at the receiver after multiple reflections may of course be used as well. The path to the receiver taken by the ray (repeated reflection, diffracted, direct) is immaterial; what really matters is the ability of the receiver to detect the ray and the individual controllability of the ray independent of other rays.

Two-dimensional and three-dimensional wireless channels are being considered in the present book. In the two-dimensional channel, the wave sources are located in one, for definiteness, horizontal plane. In such a case, the fields are independent of the vertical coordinate z. In the three-dimensional wireless channel the position of the wave sources is given by two spatial angles θ and φ. θ is the angle between the direction of the wave arrival and the vertical (axis z). φ is the azimuth angle. In this case the receive area is specified as a limited area in a three-dimensional space.

So, one of the problems being solved in this book is calculation of the limit capacity of the multipath wireless channel with the receive area specified in space. Transition has been made from the MEA system with a multiunit antenna to the potential antenna, i.e. the antenna with an unlimited number of elements confined to a limited area of space. The optimal number of spatial subchannels is thus dictated by the size of the receiving area rather than by the number of the antenna elements as in MEA systems.

Sometimes the number of spatial subchannels N can be readily estimated beforehand. If all N subchannels are identical, then the capacity C_N is calculated by the formula that follows from (1.2)

$$C_N = N \log\left(1 + \frac{SNR}{N}\right) \tag{1.4}$$

Here allowances are made for the N-fold drop in power in each subchannel. It is pertinent to note that relationship (1.1) follows from (1.4) if N indefinitely increases.

However, in the majority of cases, a tentative evaluation of the number of subchannels is difficult. The optimal amount of spatial subchannels has to be calculated. In this work, we provide the necessary computational relationships and numerous SNR dependency plots for the optimal number of spatial subchannels and the limit capacity in MIMO communication systems.

One of the interesting and significant results obtained in this avenue of research is that a receiver with large spatial size is not always a prerequisite for ensuring the optimal amount of spatial subchannels. In principle, construction of a portable receiver with the capacity approaching the limit one (1.1) is feasible.

Solving the limit throughput problem means in essence an inquiry into the deterministic wireless channel. It is assumed here that the characteristics of the radio channel are measured and known both at the transmitter and at the receiver. All the wave sources (rays) are assumed to be controlled by us independently by bringing the complex amplitudes into agreement with the characteristics of the wireless channel. In other words, we assume that all technical conditions for achieving the maximum capacity have been established.

1.3 Non-ray Statistical Models of Multipath Wireless Channels

The construction of statistical models of multipath wireless channels is the second problem addressed in this book. In contrast to the previous problem, it is assumed here that simultaneous control over different wave sources is impossible, since they are statistically independent and random. Our goal is to specify the electromagnetic field created by these sources in a limited area of space.

This field is observed in the receive area prior to measuring the channel characteristics and before exercising control over the sources. This field may also represent the model of a noise field, i.e., the field created by the random sources that pose interference for the receiver in question.

The model of a random wireless channel permits evaluation of the viability of a communication system and an appraisal of its characteristics at the design stage. The classic approach to modeling the multipath wireless channel has been described in [1-3]. It involves summing up the plane wave fields (rays) originating from a variety of random sources.

While building a statistical model of the wireless channel, paradoxical though it may seem, it sometimes pays to abandon the concept of paths and explore a different approach. This approach stems from the fact that the electromagnetic field in the receive area presents a solution of Maxwell's equations in free space. The path or the plane wave represents a solution of Maxwell's equations by the partition method in Cartesian coordinates. Employment of coordinates other than Cartesian yields different solutions of Maxwell's equations, which can be used for modeling random fields.

We use polar solutions of the equations for modeling two-dimensional fields in this book. In modeling three-dimensional fields, the solutions to Maxwell's equations in spherical coordinates represented by spherical harmonics are used. A random field is represented by a sum of well-known basis functions with random coefficients.

By way of example, we demonstrate the mathematical expression for the E_z vertical component of a random electric field of a two-dimensional omnidirectional wireless channel.

$$E_z(r, \varphi) = E_0 \sum_{n=-N}^{N} i^n \alpha_n J_n\left(\frac{2\pi}{\lambda_w} r\right) \cdot e^{in\varphi} \tag{1.5}$$

In (1.5) E_0 denotes the mean-square value of the electric field vertical component; r and φ designate the plane polar coordinates with datum origin in the center of the receive area, J_n denotes the Bessel function of the first kind, λ_w denotes wavelength, and α_n represents statistically independent complex random values with zero mean and unit variance. The $(2N+1)$ number of coefficients α_n defines the size of the area to which the model (1.4) applies.

The expression for the field vertical component H_z is similar to (1.5), it differs only in that the β_n coefficients are used in lieu of α_n. All the rest of the electric and magnetic field components are determined through α_n and β_n.

The model of a 3-D wireless channel is built in a similar way, but through use of other basis functions. The proposed models make allowances for the Doppler effect brought about by the motion of the receiver, but in a way different from that typical of the classic model.

An electromagnetic field is generated in the model, the complex amplitudes of which are randomly space-variant, but time-invariant. When the receiver is in motion, the complex amplitude at its output becomes time-variant due to changes in the external field. The spectrum of this varying signal represents the Doppler spectrum.

There are several merits to the proposed models. The models determine the coordinate dependence of all spatial components of the electric and magnetic fields in the receive area. This is why they are well suited for the evaluation of spatially treated systems.

The proposed models determine the electromagnetic field and are not tied to an antenna of any particular design. These models may be combined with various antenna models for selecting the most adequate design.

Besides, the models are relatively simple with the random coefficients of the models having uncomplicated statistical properties. Use of the models shows promise for developing effective simulations for evaluating the working capacity and characteristics of the would-be radio communication systems.

1.4 Summary

Two issues of interest for specialists in communication systems and antenna equipment have been stated for further consideration in the book.

First, the problem of estimating the limit capacity has been set forth for a hypothetical communication system, in which the potential receive antenna extracts the entirety of information from the electromagnetic field in a specified limited receiving area.

Secondly, a non-ray approach to statistical modeling the random multipath wireless channel has been suggested. The proposed approach is based on use of the spherical harmonic fields as basis functions in modeling.

1.4 Summary

2 Limit Capacity and Statistical Models of Wireless Channels

This chapter is intended for the readers who, being reluctant to follow the intricacies of mathematical transformations, are prepared to take the computational equations presented in the book on trust. This chapter serves the reference purposes and contains major design formulae. Using the calculation apparatus presented here the reader will be able to estimate the limit capacity and the required optimal quantity of spatial subchannels all by himself. In doing so, the reader can vary the number of radiation sources, their spatial arrangement, the shape of the receiving area and other initial parameters.

The limit capacity formulae for the two-dimensional wireless channel are presented in Sect. 2.1. The three-dimensional wireless channel is analyzed in Sect. 2.2. In addition, the chapter contains computational relationships for building statistical models of the two-dimensional and three-dimensional omnidirectional wireless channels. We dwell upon aleatory two-dimensional and three-dimensional channels in Sect. 2.3 and Sect. 2.4 respectively.

Each of the four sections of this chapter may be read independently of the others. We shall have more to say about how all the results have been derived further in the book. Therefore, the inquisitive reader who is not inclined to take things for gospel may skip Chap. 2 and proceed to Chap. 3.

2.1 Limit Capacity of the Two-Dimensional Wireless Channel

In the two-dimensional wireless channel, fields are independent of one of the spatial coordinates (z for definiteness). The electromagnetic field of the two-dimensional wireless channel is defined through the vertical component of the electric field E_z (E–waves) and the vertical component of the magnetic field H_z (H–waves). It is assumed that these components can be observed in a limited region in the plane (the receiving area).

The calculation formulae imply identical contribution of the waves of both types to the limit capacity. This is why a reduced matrix, representing the waves of one type only, is employed during computations. After calculating the **H** matrix of the wireless channel the eigenvalues are obtained. In estimating the capacity from the eigenvalues a water-filling algorithm is applied. The water-filling algorithm was discovered by C. Shannon and was first published in the work [14], which also contained Shannon's equation (1.2). In all subsequent equations we employ the ancillary parameter W. We may refer to this parameter as "water level". The variation range of the parameter W should be selected so that the W-dependent SNR value varies within the desired bounds.

2.1.1 Limit Capacity Estimation for a Circular Receiving Area

Basic data:
- M – number of the wave sources of one type (E-type);
- φ_m – angular coordinates of the sources ($m = 0,1,...M\text{-}1$);
- r – radius of the receiving area in the wavelength units;
- N – number of spatial harmonics of the same type taken into account;
- n – running number of the spatial harmonic ($n = 0,1,...N\text{-}1$).

It is assumed that $N \leq M$;

Calculation formulae:

Calculation of channel matrix:
Calculation of the Bessel function indices $n1(n)$.

$$n1(n) = \begin{cases} n/2 ....n = even \\ -(n+1)/2 ......n = odd \end{cases} \tag{2.1}$$

Equation (2.1) turns the even values of n into positive indices n_1, and the odd ones-into negative.

The elements of the $H_{m,n}$ channel matrix are derived from

$$H_{n,m} = ((-1)^n i)^{n1(n)} J_{|n1(n)|}(2\pi r) e^{-in1(n)\varphi m} \tag{2.2}$$

In (2.2) $J_n(x)$ denotes the Bessel function of the first kind.

Calculation of the eigenvalues of the $\mathbf{H \cdot H^+}$ *matrix*:
$\mathbf{H^+}$ denotes the Hermitian conjugate matrix (i.e., the transposed and adjoint matrix) to matrix $\mathbf{H}$. The eigenvalues of the $\mathbf{H \cdot H^+}$ matrix are real and nonnegative. Let us designate them as eigenvector λ.

$$\lambda = eigenvals(\mathbf{HH^+}) \tag{2.3}$$

All the eigenvalues are assumed to be arranged in descending order and normalized to their maximum 1.

$$\lambda_0 = 1 \geq \lambda_1 \geq \lambda_2 \geq ... \geq \lambda_{N-1} \tag{2.4}$$

Calculation of the channel characteristics:

In all subsequent formulae we employ the ancillary parameter W, the water level. The variation range of the parameter W should be selected so that the W-dependent SNR value varies within the desired bounds.

The optimal number of spatial subchannels N_c is determined by the number of n's, for which the product $W\lambda_n$ is greater than 1.

$$N_C(W) = 2 \sum_{n=0}^{N-1} 1(for \ldots W\lambda_n > 1) \tag{2.5}$$

The multiplier 2 allows for presence of the waves of the second type.
The capacity per unit bandwidth $C(W)$ is given by:

$$C(W) = 2 \sum_{n=0}^{\frac{N_C(W)}{2}-1} \log(W\lambda_n) \tag{2.6}$$

The signal-to-noise ratio $SN(W)$ should be calculated by:

$$SN(W) = N_c W - 2 \sum_{n=0}^{\frac{N_C(W)}{2}-1} \frac{1}{\lambda_n} \tag{2.7}$$

In what follows instead of $SN(W)$, we shall use the convention of $SNR(W)$ in decibels.

$$SNR(W) = 10 \lg SN(W) \tag{2.8}$$

Equations (2.8) and (2.6) present the sought-for SNR-dependence of the capacity per unit bandwidth C in a parametric form. The SNR-dependence of the optimal amount of spatial subchannels N_c is given by relationships (2.8) and (2.5).

The throughput with the infinite number of sources M and infinitely large quantity of spatial harmonics N will be named the limit capacity of the wireless channel. In finding the numerical value of the limit capacity, M and N should be increased until their further augment produces no tangible impact on performance.

With a small receiving area ($r \ll \lambda_w$) the optimal number of spatial subchannels is 2, and only discrimination by the wave-type is possible (polarization diversity). In such a case the limit capacity is evaluated by the formula (1.3), if $N = 2$ in it. That is

$$C = 2 \log\left(1 + \frac{SNR}{2}\right) \tag{2.9}$$

2.1.2 Evaluation of the Limit Capacity of the Two-Dimensional Wireless Channel with a Receiving Area of an Arbitrary Shape

Basic data:
- M – number of the wave sources of one type (E-type);
- φ_m – angular coordinates of the sources ($m = 0,1,...M$);
- N – the number of the points along the area boundary, where the vertical component of the electric and magnetic fields can be detected;
- n – running number of the point ($n = 0,1,...N\text{-}1$);
- $x_n,\ y_n$ – Cartesian coordinates of the field registration points in meters

- k–wave number, related to the wavelength through the equation $k = 2\pi/\lambda_w$. We assume that $N \geq M$.

Calculation formulae:

Calculation of channel matrix:
Computation of the elements of the $H_{m,n}$ channel matrix

$$H_{n,m} = e^{ik(x_n \cos\varphi_m + y_n \sin\varphi_m)} \tag{2.10}$$

The subsequent formulae in this subsection are replicas of the equations of Subsect. 2.1.1 with m substituted for n and M for N.

Calculation of the eigenvalues of the $\mathbf{H}^+\cdot\mathbf{H}$ matrix:
$\mathbf{H}^+$ denotes the Hermitian conjugate matrix (i.e. the transposed and adjoint matrix) to matrix $\mathbf{H}$. The eigenvalues of the $\mathbf{H}^+\cdot\mathbf{H}$ matrix are real and nonnegative. Let us designate them as eigenvector λ.

$$\lambda = eigenvals(\mathbf{H}^+\mathbf{H}) \tag{2.11}$$

All the eigenvalues are assumed to be arranged in descending order and normalized to maximum eigenvalue 1.

$$\lambda_0 = 1 \geq \lambda_1 \geq \lambda_2 \geq ... \geq \lambda_{M-1} \tag{2.12}$$

Estimation of the channel characteristics:
In all subsequent formulae we employ the ancillary parameter W, the water level. The variation range of the parameter W should be selected so that the W-dependent SNR value varies within the desired bounds.

The optimal number of spatial subchannels N_c is determined by the number of m's, for which the product $W\lambda_m$ is greater than 1.

$$N_C(W) = 2 \sum_{m=0}^{M-1} 1(for...W\lambda_m > 1) \tag{2.13}$$

The multiplier 2 allows for presence of the waves of the second type.
The capacity per unit bandwidth $C(W)$ is given by

$$C(W) = 2 \sum_{m=0}^{\frac{N_C(W)}{2}-1} \log(W\lambda_m) \tag{2.14}$$

The signal-to-noise ratio $SN(W)$ should be calculated by

$$SN(W) = N_cW - 2 \sum_{m=0}^{\frac{N_C(W)}{2}-1} \frac{1}{\lambda_m} \tag{2.15}$$

In what follows instead of $SN(W)$, we shall use the convention of $SNR(W)$ in decibels.

$$SNR(W) = 10\lg SN(W) \tag{2.16}$$

Equations (2.16) and (2.14) present the sought-for SNR-dependence of the capacity per unit bandwidth C in a parametric form. The SNR-dependence of the optimal amount of spatial subchannels N_c is given by relationships (2.16) and (2.13).

The throughput with the infinite number of sources M and infinitely large quantity of the registration points N will be named the limit capacity of the wireless channel. In finding the numerical value of the limit capacity, M and N are increased so that their further increment produces no tangible effect on performance.

2.2 Limit Capacity of the Three-Dimensional Wireless Channel

In a three-dimensional wireless channel position of wave sources is specified by two angular coordinates. These are angle θ_m between the vertical and the direction to the wave source, and azimuth angle φ_m. Another characteristic of the wave propagating from the source is polarization. Two types of polarized waves are recognized, the vertically polarized waves (the vertical component of the electric field is nonzero, while the vertical component of the magnetic field equals zero) and the horizontally polarized waves (the vertical component of the magnetic field is nonzero, while the vertical component of the electric field equals zero).

The limit capacity estimation technique for the 3-D channel differs from the 2-D channel limit capacity arithmetic in terms of the distinct way of writing the matrix **H**. After calculating the **H** matrix of the wireless channel the method of determining the eigenvalues and employment of the Shannon water-filling algorithm remain the same as with the two-dimensional channel.

2.2.1 Limit Capacity Estimation for a Spherical Receiving Area

Basic data:
- M – number of wave sources of the same polarization;
- θ_m, φ_m–angular coordinates of the sources ($m = 0,1,...M\text{-}1$);

- r – radius of the spherical receiving area in terms of the wavelength;
- N – the number of spherical harmonics of the same type taken into consideration;
- n – running number of the harmonic $(n = 0,1,...N\text{-}1)$.
  We assume that $N \leq M$.

Calculation formulae:

Calculation of channel matrix:
Compute indices $n1(n)$ and $n2(n)$ of the spherical harmonics

$$n1(n) = \left[\sqrt{n+1}\right] \tag{2.17}$$

$$n2(n) = n+1-n1(n)-n1^2(n)$$

Here the brackets $[x]$ denote the integer part of x.

The **H** matrix of the wireless channel may be conveniently written with the help of the sub-arrays.

$$\mathbf{H} = \begin{bmatrix} \mathbf{H0} & -\mathbf{H1} \\ \mathbf{H1} & \mathbf{H0} \end{bmatrix} \tag{2.18}$$

Sub-arrays **H0** and **H1** are of the $N \times M$ size. Their elements can be calculated by the following formulae:

$$H0_{n,m} = i^{n1(n)-1} j_{n1(n)}(2\pi r) \frac{d(PN_{n1(n)}^{(n2(n))}(\cos\theta))}{d\theta}\bigg|_{\theta=\theta_m} e^{-in2(n)\varphi_m} \tag{2.19a}$$

$$H1_{n,m} = i^{n1(n)} j_{n1(n)}(2\pi r) \frac{n_2(n)PN_{n1(n)}^{(n2(n))}(\cos\theta_m)}{\sin\theta_m} e^{-in2(n)\varphi_m} \tag{2.19b}$$

In the expressions of (2.19) $PN_n^{(m)}(\theta)$ is the normalized associated Legendre function. The normalization by transmission power employed in this book differs from the conventional one. The normalized associated Legendre function is related to the function $P_n^{(m)}(\theta)$ by

$$PN_n^{(m)}(\theta) = \sqrt{\frac{1}{2\pi} \frac{2n+1}{n(n+1)} \frac{(n-m)!}{(n+m)!}} P_n^{(m)}(\theta) \qquad (2.20)$$

The associated Legendre function $P_n^{(m)}(\theta)$ for $m \geq 0$ can be worked out from

$$P_n^{(m)}(\theta) = \left(\cos\frac{\theta}{2}\right)^m \sum_{l=\max(m,0)}^{n} \frac{(-1)^l (n+l)!}{(n-l)!(l-m)!l!} \left(\sin\frac{\theta}{2}\right)^{2l-m} \qquad (2.21a)$$

In (2.21a) $\max(m,0)$ denotes the greatest of the two values m and 0. Notice that the $P_n^{(m)}(\theta)$ functions should be held equal to zero, if $n < m$.

For $m < 0$ the normalized associated Legendre function and its derivative can be calculated by the following formulae:

$$PN_n^{(-m)}(\theta) = (-1)^m PN_n^{(m)}(\theta) \qquad (2.21b)$$

$$\frac{d\left(PN_n^{(-m)}(\theta)\right)}{d\theta} = (-1)^m \frac{d\left(PN_n^{(m)}(\theta)\right)}{d\theta} \qquad (2.21c)$$

In (2.19) the spherical Bessel function is expressed by $j_n(x)$. To calculate it use the following formula

$$j_n(x) = \mathrm{Re}\left[i^{n+1} \frac{e^{-ix}}{x} \sum_{l=0}^{n} \frac{(n+l)!}{l!(n-l)!} (2ix)^{-l} \right] \qquad (2.22a)$$

In (2.22a) Re denotes the real part. However conveniently compact, (2.22a) may yield considerable error during computations if n is large and x is small. For this reason the following expression may happen to become the formula of choice for finding $j_n(x)$

$$j_n(x) = \frac{x^n}{1 \cdot 3 \cdot 5 \cdot \ldots \cdot (2n+1)} \left(1 - \frac{\frac{x^2}{2}}{1!(2n+3)} + \frac{\left(\frac{x^2}{2}\right)^2}{2!(2n+3)(2n+5)} - \ldots \right) \qquad (2.22b)$$

After the $2N \times 2M$ matrix $\mathbf{H}$ has been worked out, calculations similar to those in Subsect. 2.1.1 are performed, though slightly modified. This modification stems from the impossibility of considering two wave types in the three-dimensional channel independently. Let us show further mathematics necessary for determining the limit capacity of a wireless channel.

Calculation of the eigenvalues of the $\mathbf{H} \cdot \mathbf{H}^+$ matrix:

$\mathbf{H}^+$denotes the Hermitian conjugate matrix (i.e. the transposed and adjoint matrix) to matrix $\mathbf{H}$. The eigenvalues of the $\mathbf{H} \cdot \mathbf{H}^+$ matrix are real and nonnegative. Let us designate their totality as eigenvector λ.

$$\lambda = eigenvals(\mathbf{HH}^{+}) \tag{2.23}$$

All the eigenvalues are assumed to be in a descending order and normalized to the maximum eigenvalue 1.

$$\lambda_0 = 1 \geq \lambda_1 \geq \lambda_2 \geq ... \geq \lambda_{2N-1} \tag{2.24}$$

Estimation of the channel characteristics:

In all subsequent formulae we employ the ancillary parameter W, the water level. The variation range of the parameter W should be selected so that the W-dependent SNR value varies within the desired bounds.

The optimal number of spatial subchannels N_c is determined by the number of l's, for which the product $W\lambda_l$ is greater than 1.

$$N_c(W) = \sum_{l=0}^{2N-1} 1(for ... W\lambda_l > 1) \tag{2.25}$$

The capacity per unit bandwidth $C(W)$ is given by

$$C(W) = \sum_{l=0}^{N_c(W)-1} \log(W\lambda_l) \tag{2.26}$$

The signal- to-noise ratio $SN(W)$ should be calculated by means of

$$SN(W) = N_c W - \sum_{l=0}^{N_c(W)-1} \frac{1}{\lambda_l} \tag{2.27}$$

In what follows instead of $SN(W)$, we shall use the convention of $SNR(W)$ in decibels.

$$SNR(W) = 10 \lg SN(W) \tag{2.28}$$

Equations (2.28) and (2.26) present the sought-for SNR-dependence of the capacity per unit bandwidth C in a parametric form. The SNR-dependence of the optimal amount of spatial subchannels N_c is given by relationships (2.28) and (2.25).

The throughput for the infinite number of sources M and infinitely large quantity of spatial harmonics N will be named the limit capacity of the wireless channel. In finding the numerical value of the limit capacity, M and N are increased so that their further increment produces no tangible performance effect.

In an omnidirectional wireless channel with a small receiving area ($r \ll \lambda_w$) the optimal number of spatial subchannels is equal to 6. In such a case the limit capacity is evaluated with formula (1.3), with the proviso that $N = 6$. That is,

$$C = 6\log\left(1 + \frac{SNR}{6}\right) \qquad (2.29)$$

Provided that SNR is not great, six spatial subchannels ensure a limit capacity per unit bandwidth approaching the limit C_∞ (1.1), attainable with the infinite number of spatial subchannels. By this means with SNR = 5 dB, formula (2.29) yields the result by a mere 20% less than the limit capacity C_∞. This result is of crucial importance; it attests to the basic feasibility of high performance wireless communication systems with small-size antennas. In an omnidirectional wireless channel with 6 spatial subchannels no spatial separation between the antenna elements is required.

2.2.2 Estimation of the Limit Capacity of the Three-Dimensional Wireless Channel with a Receiving Area of an Arbitrary Shape

Basic data:
- M – number of wave sources of the same polarization;
- θm, φm – angular coordinates of the sources $(m = 0,1,...M\text{-}1)$;
- N – the number of points along the area boundary, where the Cartesian components of the electric and magnetic fields are picked up;
- n – running number of the point $(n = 0,1,...N\text{-}1)$;
- x_n, y_n, z_n – Cartesian coordinates of the field pickup points in meters;
- k – wave number, related to the wavelength by the equation $k = 2\pi/\lambda_w$.
  We assume that $N \geq M$.

Calculation formulae:
Calculation of channel matrix:
The **H** matrix of the wireless channel may be conveniently written with the help of the sub-arrays.

$$\mathbf{H} = \begin{bmatrix} \mathbf{HE}_x^v & \mathbf{HE}_x^h \\ \mathbf{HE}_y^v & \mathbf{HE}_y^h \\ \mathbf{HE}_z^v & \mathbf{HE}_z^h \\ \mathbf{HH}_x^v & \mathbf{HH}_x^h \\ \mathbf{HH}_y^v & \mathbf{HH}_y^h \\ \mathbf{HH}_z^v & \mathbf{HH}_z^h \end{bmatrix} \qquad (2.30)$$

All the sub-arrays in (2.30) are of size $N \times M$. Their elements are computed by the following formulae:

$$HE_{xn,m}^v = \cos\theta_m \cos\varphi_m e^{i(\mathbf{k}\cdot\mathbf{r})} \qquad (2.31\text{av})$$

$$HE^{v}_{yn,m} = \cos\theta_m \sin\varphi_m e^{i(\mathbf{k}\cdot\mathbf{r})} \qquad (2.31\text{bv})$$

$$HE^{v}_{zn,m} = -\sin\theta_m e^{i(\mathbf{k}\cdot\mathbf{r})} \qquad (2.31\text{cv})$$

$$HH^{v}_{xn,m} = \sin\varphi_m e^{i(\mathbf{k}\cdot\mathbf{r})} \qquad (2.31\text{dv})$$

$$HH^{v}_{yn,m} = -\cos\varphi_m e^{i(\mathbf{k}\cdot\mathbf{r})} \qquad (2.31\text{ev})$$

$$HH^{v}_{zn,m} = 0 \qquad (2.31\text{fv})$$

$$HE^{h}_{xn,m} = -\sin\varphi_m e^{i(\mathbf{k}\cdot\mathbf{r})} \qquad (2.31\text{ah})$$

$$HE^{h}_{yn,m} = \cos\varphi_m e^{i(\mathbf{k}\cdot\mathbf{r})} \qquad (2.31\text{bh})$$

$$HE^{h}_{zn,m} = 0 \qquad (2.31\text{ch})$$

$$HH^{h}_{xn,m} = \cos\theta_m \cos\varphi_m e^{i(\mathbf{k}\cdot\mathbf{r})} \qquad (2.31\text{dh})$$

$$HH^{h}_{yn,m} = \cos\theta_m \sin\varphi_m e^{i(\mathbf{k}\cdot\mathbf{r})} \qquad (2.31\text{eh})$$

$$HH^{h}_{zn,m} = -\sin\theta_m e^{i(\mathbf{k}\cdot\mathbf{r})} \qquad (2.31\text{dh})$$

the scalar product $(\mathbf{k}\cdot\mathbf{r})$ included in expressions (2.31) is obtained from

$$(\mathbf{k}\cdot\mathbf{r}) = kx_n \sin\theta_m \cos\varphi_m + ky_n \sin\theta_m \sin\varphi_m + kz_n \cos\theta_m \qquad (2.32)$$

Other equations of this Subsection are similar to those of Subsect. 2.2.1.

Calculation of the eigenvalues of the $\mathbf{H}^{+}\cdot\mathbf{H}$ *matrix*:
$\mathbf{H}^{+}$denotes the Hermitian conjugate matrix (i.e. the transposed and adjoint matrix) to matrix $\mathbf{H}$. The eigenvalues of the $\mathbf{H}^{+}\cdot\mathbf{H}$ matrix are real and nonnegative. Let us designate their totality as eigenvector λ.

$$\lambda = eigenvals(\mathbf{H}^{+}\mathbf{H}) \qquad (2.33)$$

All the eigenvalues are assumed to be in descending order and normalized to the maximum eigenvalue 1.

$$\lambda_0 = 1 \geq \lambda_1 \geq \lambda_2 \geq \ldots \geq \lambda_{2M-1} \tag{2.34}$$

Estimation of channel characteristics:

In all subsequent formulae we employ the ancillary parameter W, the water level. The variation range of the parameter W should be selected so that the W-dependent SNR value varies within the desired bounds.

The optimal number of spatial subchannels N_c is determined by the number of l's, for which the product $W\lambda_l$ is greater than 1.

$$N_c(W) = \sum_{l=0}^{2M-1} 1(for \ldots W\lambda_l > 1) \tag{2.35}$$

The capacity per unit bandwidth $C(W)$ is given by

$$C(W) = \sum_{l=0}^{N_C(W)-1} \log(W\lambda_l) \tag{2.36}$$

The signal-to-noise ratio $SN(W)$ should be calculated by means of

$$SN(W) = N_c W - \sum_{l=0}^{N_C(W)-1} \frac{1}{\lambda_l} \tag{2.37}$$

In what follows instead of $SN(W)$, we shall use the convention of $SNR(W)$ in decibels.

$$SNR(W) = 10 \lg SN(W) \tag{2.38}$$

Equations (2.38) and (2.36) present the sought-for SNR-dependence of the capacity per unit bandwidth C in a parametric form. The SNR-dependence of the optimal number of spatial subchannels N_c is given by relationships (2.38) and (2.35).

The throughput for the infinite number of sources M and infinitely large quantity of the registration points N will be named the limit capacity of the wireless channel. In finding the numerical value of the limit capacity, M and N are increased so that their further increment produces no tangible effect on performance.

2.3 Statistical Model of the Two-Dimensional Omnidirectional Wireless Channel

A statistical model of the 2-D wireless channel permits generation of random fields. The number of field-generating sources is assumed to be numerous, while their angular distribution φ is held to be uniform.

All the components of the electric and magnetic field in the neighborhood of the coordinate origin are determined by a set of random coefficients α_n and β_n. For the omnidirectional channel all coefficients are independent complex random normal quantities with identical variances and zero mean. For a normalized field with unit variance of the vertical field component E_z the variance of each coefficient is 1 (the variance of the real and imaginary part being 0.5 each).

Model parameters:
- N – maximum index of the Bessel function used in calculation. The quantity of random coefficients is connected with the size of N. Their amount is $2N+1$ for α_n as well as for β_n ($n = -N,-(N-1)...(N-1),N$) alike. An increase of N causes expansion of the region of space where the model relationships hold true.
- k is the wave number, related to the wavelength by the equation $k = 2\pi/\lambda_w$.

Calculation of electromagnetic field:

Projections of the electric and magnetic fields in cylindrical coordinates (r, φ, z) are given by:

$$E_r = \frac{1}{kr}\sqrt{\frac{\mu}{\varepsilon}}\sum_{n=-N}^{N} i^{n-2} n\beta_n J_n(kr)e^{in\varphi} \tag{2.39a}$$

$$E_\varphi = \sqrt{\frac{\mu}{\varepsilon}}\sum_{n=-N}^{N} i^{n-1}\beta_n J_n^1(kr)e^{in\varphi} \tag{2.39b}$$

$$E_z = \sum_{n=-N}^{N} i^n \alpha_n J_n(kr)e^{in\varphi} \tag{2.39c}$$

$$H_r = \frac{1}{kr}\sqrt{\frac{\varepsilon}{\mu}}\sum_{n=-N}^{N} i^n n\alpha_n J_n(kr)e^{in\varphi} \tag{2.39d}$$

$$H_\varphi = \sqrt{\frac{\varepsilon}{\mu}}\sum_{n=-N}^{N} i^{n+1}\alpha_n J_n^1(kr)e^{in\varphi} \tag{2.39e}$$

$$H_z = \sum_{n=-N}^{N} i^n \beta_n J_n(kr)e^{in\varphi} \tag{2.39f}$$

In the formulae (2.39) $J_n(x)$ denotes the Bessel function of the first kind. $J'_n(x)$ is used to designate the derivative of the Bessel function.

The quantity $(\mu/\varepsilon)^{1/2}$ found in formulae (2.39) deserves particular attention. This quantity is the characteristic impedance of free space, which equals 120π [Ω] in the SI system. However, during a numerical field modeling it is advantageous to take it to be equal to 1.

$$\sqrt{\frac{\mu}{\varepsilon}} = 1 \qquad (2.40)$$

The equality (2.40) means that the sensor picking up the magnetic field of a plane wave and the sensor picking up its electric field provide identical response. In other words picking up the plane wave by its magnetic field is tantamount to picking it up by the electric field. It is under this assumption that the coefficients α_n and β_n have identical variances.

The interrelation of the formulae (2.39) with the ray-oriented approach is described in Sect. 4.4. Consideration of wireless channels other than the omnidirectional one is presented in the same Section.

2.4 Statistical Model of the Three-Dimensional Omnidirectional Wireless Channel

A statistical model of the 3-D wireless channel permits generation of random fields dependent on three spatial coordinates. The number of the field-producing sources is assumed to be large. It is further assumed that the sources are uniformly distributed over the surface of a distant sphere.

All the electric and magnetic field components in the region of the coordinate origin are determined by a set of random coefficients $\alpha_{n,m}$ and $\beta_{n,m}$. In the case of the omnidirectional wireless channel, all coefficients are independent complex normal random values with zero mean and identical variances. For a normalized field with the unity variance vertical component E_z the variance of each coefficient equals 3π (with the variances of the real and imaginary parts being 1.5π each).

Model parameters:
- N identifies the maximal index of the spherical Bessel function used in the calculation. The running index n takes on values 1, 2, ..., N. For each value of n there are $2n + 1$ values of m ($m = -n$, $-(n-1)$, ..., $(n-1)$, n). The numerical value of N is connected with the number of the random coefficients. Their number is N^2-1 for α and for β alike. Increase of N causes expansion of the region of space where the model relationships hold true.
- k denotes the wave number, related to the wavelength by the equation $k = 2\pi/\lambda_w$.

Calculation of electromagnetic field:

Let us now write the calculation formulae for projections of the electric and magnetic fields in spherical coordinates (r, θ, φ). These random fields are represented by sums of the fields of the spherical harmonics with random coefficients.

$$E_r = \sum_{n=1}^{N} \sum_{m=-n}^{n} \alpha_{n,m} \left(\frac{d^2 [krj_n(kr)]}{d(kr)^2} + krj_n(kr) \right) PN_n^{(m)}(\theta) e^{im\varphi} \tag{2.41a}$$

$$E_\theta = \frac{1}{kr} \sum_{n=1}^{N} \sum_{m=-n}^{n} \alpha_{n,m} \frac{d[krj_n(kr)]}{d(kr)} \frac{d(PN_n^{(m)}(\theta))}{d\theta} e^{im\varphi} +$$
$$+ \frac{1}{\sin\theta} \sqrt{\frac{\mu}{\varepsilon}} \sum_{n=1}^{N} \sum_{m=-n}^{n} \beta_{n,m} mj_n(kr) PN_n^{(m)}(\theta) e^{im\varphi} \tag{2.41b}$$

$$E_\varphi = \frac{i}{kr\sin\theta} \sum_{n=1}^{N} \sum_{m=-n}^{n} \alpha_{n,m} m \frac{d[krj_n(kr)]}{d(kr)} PN_n^{(m)}(\theta) e^{im\varphi} +$$
$$+ i\sqrt{\frac{\mu}{\varepsilon}} \sum_{n=1}^{N} \sum_{m=-n}^{n} \beta_{n,m} j_n(kr) \frac{d(PN_n^{(m)}(\theta))}{d\theta} e^{im\varphi} \tag{2.41c}$$

$$H_r = \sum_{n=1}^{N} \sum_{m=-n}^{n} \beta_{n,m} \left(\frac{d^2 [krj_n(kr)]}{d(kr)^2} + krj_n(kr) \right) PN_n^{(m)}(\theta) e^{im\varphi} \tag{2.41d}$$

$$H_\theta = \frac{1}{kr} \sum_{n=1}^{N} \sum_{m=-n}^{n} \beta_{n,m} \frac{d[krj_n(kr)]}{d(kr)} \frac{d(PN_n^{(m)}(\theta))}{d\theta} e^{im\varphi} -$$
$$- \frac{1}{\sin\theta} \sqrt{\frac{\varepsilon}{\mu}} \sum_{n=1}^{N} \sum_{m=-n}^{n} \alpha_{n,m} mj_n(kr) PN_n^{(m)}(\theta) e^{im\varphi} \tag{2.41e}$$

$$H_\varphi = \frac{i}{kr\sin\theta} \sum_{n=1}^{N} \sum_{m=-n}^{n} \beta_{n,m} m \frac{d[krj_n(kr)]}{d(kr)} PN_n^{(m)}(\theta) e^{im\varphi} -$$
$$- i\sqrt{\frac{\varepsilon}{\mu}} \sum_{n=1}^{N} \sum_{m=-n}^{n} \alpha_{n,m} j_n(kr) \frac{d(PN_n^{(m)}(\theta))}{d\theta} e^{im\varphi} \tag{2.41f}$$

In the formulae (2.41), $PN_n^{(m)}(\theta)$ and $j_n(x)$ denote the same functions that were used in writing formulae (2.19). $PN_n^{(m)}(\theta)$ is the normalized associated Legendre function. The normalization by the radiation power employed in this book differs from the one commonly used. The normalized associated Legendre function is related to the function $P_n^{(m)}(\theta)$ through the relationship

$$PN_n^{(m)}(\theta) = \sqrt{\frac{1}{2\pi} \frac{2n+1}{n(n+1)} \frac{(n-m)!}{(n+m)!}} P_n^{(m)}(\theta) \tag{2.42}$$

The associated Legendre function $P_n^{(m)}(\theta)$ for $m \geq 0$ may be calculated by

$$P_n^{(m)}(\theta) = \left(\cos\frac{\theta}{2} \right)^m \sum_{l=\max(m,0)}^{n} \frac{(-1)^l (n+l)!}{(n-l)!(l-m)!l!} \left(\sin\frac{\theta}{2} \right)^{2l-m} \tag{2.43a}$$

In (2.43a) $\max(m,0)$ denotes the greatest of the two values m and 0. It is worth-while noting that the $P_n^{(m)}(\theta)$ functions should be held equal to zero, if $n < m$.

For m < 0 the normalized associated Legendre function and its derivative can be calculated by the following formulae:

$$PN_n^{(-m)}(\theta) = (-1)^m PN_n^{(m)}(\theta) \tag{2.43b}$$

$$\frac{d\left(PN_n^{(-m)}(\theta)\right)}{d\theta} = (-1)^m \frac{d\left(PN_n^{(m)}(\theta)\right)}{d\theta} \tag{2/43c}$$

$j_n(x)$ in (2.41) designates the spherical Bessel function. It can be found by

$$j_n(x) = \mathrm{Re}\left(i^{n+1} \frac{e^{-ix}}{x} \sum_{l=0}^{n} \frac{(n+l)!}{l!(n-l)!} (2ix)^{-l} \right) \tag{2.44a}$$

In (2.44a) Re denotes the real part. Though compact (2.44a) may produce material error during computation if n is big and x is small. For this reason preference may be given to finding $j_n(x)$ with the help of

$$j_n(x) = \frac{x^n}{1 \cdot 3 \cdot 5 \cdot \ldots \cdot (2n+1)} \left(1 - \frac{\frac{x^2}{2}}{1!(2n+3)} + \frac{\left(\frac{x^2}{2}\right)^2}{2!(2n+3)(2n+5)} - \ldots \right) \tag{2.44b}$$

The quantity $(\mu/\varepsilon)^{1/2}$ present in the formulae (2.41) is worth special mention. This value is the characteristic impedance of free space that equals 120π [Ohm] in the SI system. However, during numerical modeling it should be assumed equal to 1.

$$\sqrt{\frac{\mu}{\varepsilon}} = 1 \tag{2.45}$$

The equality (2.45) means that the sensor picking up the magnetic field of a plane wave and the sensor picking up its electric field provide identical response. In other words, picking up the plane wave by its magnetic field is tantamount to picking it up by the electric field. It is under this assumption that the coefficients $\alpha_{n,m}$ and $\beta_{n,m}$ have identical variances.

However cumbersome, the formulae (2.41) create no problem in computer-aided calculations. Notice that the expressions are helpful in finding all the components of the electric and magnetic fields in the neighborhood of the coordinate origin. A more detailed consideration of the three-dimension channel model is given in Chap. 8.

2.5 Summary

Chapter 2 presents the simplest of the obtained computational relationships without deduction. A more detailed presentation of the findings is given in the subsequent chapters of this book.

The limit capacity of the two-dimensional wireless channel briefly tackled in 2.1 is considered in Chap. 4, excluding Sect. 4.4. Chap. 5, 6 and 7 are dedicated to analyzing the limit capacity of the three-dimensional wireless channel, described in 2.2.

The detailed description of the 2-D channel statistical model overviewed in Sect. 2.3 is presented in Sect. 4.4. The statistical model of the three-dimension wireless channel, outlined in 2.4, is examined in Chap. 8. Previous familiarity with the content of Chap. 5 will provide a better understanding of Chap. 8.

3 Capacity of Multiple Input Multiple Output Communication Systems

This chapter provides the insights necessary for estimating the capacity of multiple input multiple output (MIMO) systems. The representation of the MIMO system is considered here as a multitude of individual and independent subchannels. A method for handling the problem of optimal power distribution with various types of interference is presented.

3.1 Decomposition of MIMO Systems

Let us consider a linear system with L inputs and M outputs which is characterized by the relationship

$$\mathbf{B}_0 = \mathbf{H} \cdot \mathbf{A}_0 + \mathbf{n}_0 \tag{3.1}$$

Here $\mathbf{A}_0$ denotes the L-size vector of the input complex amplitudes, $\mathbf{B}_0$ stands for the M-size vector of the output complex amplitudes, $\mathbf{H}$ signifies the $M \times L$ size matrix of complex gains. For the sake of definiteness we assume that the number of inputs is less than the number of outputs ($L < M$), $\mathbf{n}_0$ denotes the M-size vector of the output noise, and the components of the noise vector $\mathbf{n}_0$ are held to be independent normal random quantities with equal variances.

When calculating the channel capacity determined by relationship (3.1), we will employ the technique described in [8]. It is based on the linear transformation, which recasts relationship (3.1) in the diagonal form. To this end singular representation of the $\mathbf{H}$ channel matrix should be used

$$\mathbf{H} = \mathbf{U} \cdot \mathbf{D} \cdot \mathbf{V}^+ \tag{3.2}$$

In (3.2) the superscript symbol $^+$ denotes a Hermitian conjugate, i.e. a complex conjugate and transposed matrix. $\mathbf{U}$ signifies a unitary matrix, that is the matrix whose product by the Hermitian conjugate yields the unity matrix $\mathbf{I}$

$$\mathbf{U} \cdot \mathbf{U}^+ = \mathbf{I}$$

The columns of the $\mathbf{U}$ matrix contain the eigenvectors of the $\mathbf{H \cdot H^+}$ matrix. Therefore, the $\mathbf{U}$ matrix satisfies the equality

$$\mathbf{H \cdot H^+ \cdot U = U \cdot D^2} \tag{3.3}$$

The $\mathbf{V}$ matrix is a unitary matrix, whose columns contain the eigenvectors of the $\mathbf{H^+H}$ matrix. It satisfies the equality

$$\mathbf{H^+ \cdot H \cdot V = V \cdot D^2} \tag{3.4}$$

The $\mathbf{U}$ and $\mathbf{V}$ matrices are linked to each other.

$$\mathbf{U = H \cdot V \cdot D^{-1}}$$

$$\mathbf{V = U^+ \cdot H \cdot D^{-1}}$$

The $\mathbf{D}$ matrix is a diagonal matrix array, containing the square roots of the eigenvalues of the $\mathbf{H \cdot H^+}$ matrices (or the $\mathbf{H^+H}$ ones).

The $\mathbf{H \cdot H^+}$ and $\mathbf{H^+H}$ matrix arrays are positively defined. Their nonzero eigenvalues are positive. The nonzero eigenvalues of the $\mathbf{H \cdot H^+}$ and $\mathbf{H^+H}$ arrays coincide. Therefore, for convenience in computations we can use the array with a smaller amount of eigenvalues (including zero ones.)

Substitution of (3.2) into (3.1) and introduction of new designations

$$\mathbf{B = U^+ \cdot B_0}$$

$$\mathbf{A = V^+ \cdot A_0}$$

permits representation of the channel as follows

$$\mathbf{B = D \cdot A + n} \tag{3.5}$$

The noise vector $\mathbf{n}$ in (3.5) is related to the noise vector $\mathbf{n_0}$ of (3.1) by the equality

$$\mathbf{n = U^+ \cdot n_0} \tag{3.6}$$

Notice that when multiplied by the $\mathbf{U^+}$ unitary matrix the statistical characteristics of the Gaussian noise components remain unchanged and their variances stay the same.

The matrix equality (3.5) can be written component-wise as

$$b_l = \sqrt{\lambda_l} \cdot a_l + n_l \tag{3.7}$$

where $l = 0, 1, 2, ..., L-1$.

Formula (3.7) presents the wireless channel in form of individual subchannels independent of each other. a_l and b_l denote the signals at the input and output of the l–th subchannel, n_l stands for the noise at the subchannel output, λ_l signifies the power gain of the l–th subchannel equal to the l-th eigenvalue of the $\mathbf{H \cdot H^+}$ matrix array (it is assumed, for the sake of definiteness, that the number of inputs is less than the number of outputs $L < M$).

It is assumed that all eigenvalues are integrated into the Λ vector, normalized and arranged in the descending order.

$$\lambda_0 = 1 \geq \lambda_1 \geq \lambda_2 \geq \ldots \geq \lambda_{L-1} \geq 0 \qquad (3.8)$$

It is further assumed that all the noise components are independent Gaussian random quantities with zero mean and variance D_l. Then, Shannon's formula can be used for calculation of the capacity per unit bandwidth (throughput per 1Hz of bandwidth)

$$C = \sum_{l=0}^{L-1} \log\left(1 + \frac{P_l \lambda_l}{D_l}\right) \qquad (3.9)$$

In (3.9) P_l denotes the power at the input of the l-th subchannel. The overall power P is held to be specified at the inputs of all the subchannels.

$$P = \sum_{l=0}^{L-1} P_l \qquad (3.10)$$

In order to use (3.9), it is necessary to solve the problem of optimal power distribution over individual subchannels subject to the equality (3.10). Solution of this problem is dependent on the nature of the noise. Here we examine three types of noise: thermal noise, external noise (interference) and the subchannel crosstalk of a MIMO system. We concern ourselves with the impact of the thermal noise and interference in Sect. 3.2, and look at the interchannel crosstalk in Sect. 3.3.

3.2 Optimal Power Distribution with Allowance for Thermal Noise and Interference

In this section it is assumed that the noise variance D_l in the l–th subchannel may be written as

$$D_l = D^T + D^I \lambda_l \qquad (3.11)$$

We conceive the first summand in (3.11) as being due to the noise at the output of the subchannel. Its intensity D^T is independent of the subchannel gain λ_l. This component will be named thermal noise, though it may not necessarily originate in heat sources.

The second summand in (3.11) is due to adverse external effects. The intensity D^I of this noise, along with the intensity of the legitimate signal varies λ_l times in the l-th channel.

Rewriting the capacity expression (3.9) in terms of (3.11) we obtain

$$C = \sum_{l=0}^{L-1} \log\left(1 + \frac{P_l \lambda_l}{D^T + D^I \lambda_l}\right) \qquad (3.12)$$

Subject to (3.10) it is necessary to find the P_l power values, for which the capacity is maximum. This problem has been solved by C. Shannon. The algorithm proposed by him was published for the first time in [14] where also Shannon's formula (1.2) appears. Let us examine the derivation of this algorithm.

In solution of the optimization problem, we will use the Lagrange method of undetermined multipliers. According to this method, in order to find the optimal P_l values one has to work out the extremum-of-function problem

$$\sum_{l=0}^{L-1} \log\left(1 + \frac{P_l \lambda_l}{D^T + D^I \lambda_l}\right) + a\left(P - \sum_{l=0}^{L-1} P_l\right)$$

given obvious limitations $P_l > 0$. Here a denotes the undetermined constant. Setting the derivatives of the above expression with respect to P_l equal to zero yields the following set of equations

$$P_l + D^I + \frac{D^T}{\lambda_l} = W \qquad (3.13)$$

Here a new constant (the "water level") is symbolized by W. It is related to the previous constant a through the equality $W = (a \cdot \ln 2)^{-1}$. Using D ($D = D^T + D^I$) to denote the overall variance of the thermal noise and interference in the subchannel with maximum gain ($\lambda_0 = 1$) and dividing by it both parts of the equation (3.13) we will write the channel capacity expression as

$$p_l = \begin{cases} w - d^I - \dfrac{d^T}{\lambda_l} \quad ... \, for \, ... \, w - d^I - \dfrac{d^T}{\lambda_l} > 0 \\[4mm] 0 \, ... \, for \, ... \, w - d^I - \dfrac{d^T}{\lambda_l} \leq 0 \end{cases} \qquad (3.14)$$

With normalized quantities introduced into (3.14):

$$p_l = \frac{P_l}{D}, \quad w = \frac{W}{D}, \quad d^T = \frac{D^T}{D}, \quad d^I = \frac{D^I}{D} \qquad (3.15)$$

The number of employed subchannels N_C is determined by the number of l values, for which the p_l values in (3.14) are positive.

$$N_C = l \ldots if \ldots p_{l-1} > 0, p_l \leq 0 \qquad (3.16)$$

The w parameter is defined by the specified overall power, i.e. by equation (3.10), which being divided by D takes the form:

$$SNR = \sum_{l=0}^{N_C-1} p_l \qquad (3.17)$$

The ratio of the overall power P to the variance D is labeled SNR. It must be emphasized that the SN ratio is taken to mean the ratio of the legitimate signal power to the noise variance at the output of the channel with maximum gain, if the entire input power enters this channel. The signal-to-noise ratio in each individual subchannel will be less than SNR.

The capacity-per-unit-bandwidth expression (3.12) in terms of (3.14) and (3.15) is transformed into

$$C = N_C \log w - \sum_{l=0}^{N_C-1} \log \left(\frac{d^T}{\lambda_l} + d^I \right) \qquad (3.18)$$

Therefore, expressions (3.14-3.18) permit calculation of the characteristics of the system with independent parallel channels. Expression (3.17), written in terms of (3.14), presents the equation for finding the w parameter for a given SNR value. The w parameter can then be used to find the optimal number of channels (3.16) and the system capacity (3.18).

While plotting the SNR-dependence curves for the optimal number of channels and capacity, the order of computations may be slightly modified. For example, it is not unreasonable to regard w as a variable. In this case the expressions (3.17) and (3.16) represent the SNR-dependence of the optimal number of channels in a parametric form. Expressions (3.18) and (3.16) provide a parametric representation of the SNR dependence of the capacity. Such an approach permits much quicker computations since there is no need to resolve equations.

Let us consider, for example, a MIMO system with ten independent subchannels, whose power gains vary linearly. (Fig. 3.1).

$$\lambda_l = 1 - \frac{l}{L} \qquad (3.19)$$

The plots for optimal power distribution over the subchannels are depicted in Fig. 3.2. Curve 1 represents absence of the interference noise ($d^I=0$, $d^T=1$), while Curve 2 accounts for the case when the interference noise prevails. The thermal noise level is -10 dB of the overall noise ($d^I=0.9$, $d^T=0.1$).

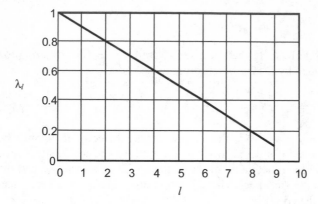

Fig. 3.1. Gain variation of subchannels

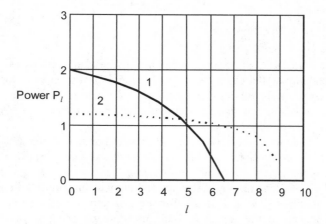

Fig. 3.2. Impact of the interference noise on power distribution over the channels. 1 – interference noise is absent ($d^I=0$, $d^T=1$). 2 – interference noise prevails ($d^I=0.9$, $d^T=0.1$)

A comparison of Curves 1 and 2 shows that when interference noise prevails, the MIMO system demonstrates a better performance than with thermal noise. With interference noise, all of the ten subchannels are used. In absence of the interference noise, the number of the employed channels drops to 7. The interference noise has been found to prevail in cellular communication systems [16]. In analyses of such systems the effect of the thermal noise is often neglected. While estimating the limit capacity the effect of the thermal noise cannot be disregarded. In this case, as is evident from expressions (3.14) and (3.16), the number of subchannels becomes equal to the number of inputs L. The value of L during field analysis (continuous coordinate functions) is indefinitely large. For this reason thermal noise will always be taken into consideration henceforth in this book. With the thermal noise taken into account, weak subchannels are not used, and the optimal number of subchannels becomes finite.

3.3 Crosstalk Effect

In this section we examine the effect of crosstalk between individual subchannels of the MIMO communication system in Sect. 3.3. According to (3.7) the individual subchannels of a MIMO system are independent only when the **H** channel matrix is known with certainty. In real communication systems the channel matrix is measured with a certain degree of error. Therefore, accurate distinction of individual subchannels becomes unfeasible; they may be linked to each other, and the signal from one subchannel leaks into another causing interference. Such interference is known as inter-channel crosstalk.

To illustrate the effect of inter-channel crosstalk on performance we assume that the power gain is the same in all individual subchannels and equals Δ.

$$K_{l1,l2} = \Delta \quad \text{with} \quad l1 \neq l2.$$

We designate the eigengain of every individual subchannel as λ_l.

$$K_{l,l} = \lambda_l$$

The crosstalk variance in the k–th subchannel may be written

$$D_{cross} = \Delta \sum_{\substack{l=0 \\ l \neq k}}^{L-1} P_l$$

The overall noise variance (the integrated one made up by thermal, interference and crosstalk noise) in the l–th subchannel may then be presented as

$$D_l = D^T + D^I \lambda_l + P\Delta - P_l\Delta \tag{3.20}$$

With the assumption that the density function of the interchannel crosstalk may be approximated with the Gaussian function, we rewrite expression (3.9) for the capacity per unit bandwidth

$$C = \sum_{l=0}^{L-1} \log\left(1 + \frac{P_l\lambda_l}{A_l - P_l\Delta}\right) \tag{3.21}$$

In this A_l denotes a P_l-independent quantity

$$A_l = D^T + D^I \lambda_l + P\Delta$$

Maximization of (3.21) with the earlier proviso of the input power invariability leads to the extremum-of-function problem

$$\sum_{l=0}^{L-1} \log\left(1 + \frac{P_l \lambda_l}{A_l - P_l \Delta}\right) + a\left(P - \sum_{l=0}^{L-1} P_l\right)$$

Differentiating the written function with respect to P_l and putting the derivative equal to zero, we obtain

$$P_l^2 - P_l \frac{A_l}{\Delta} \frac{\lambda_l - 2\Delta}{\lambda_l - \Delta} - \frac{A_l}{\Delta} \frac{A_l - W\lambda_l}{\lambda_l - \Delta} = 0$$

where W is a constant related to the earlier constant a through the relationship $W = (a \cdot \ln 2)^{-1}$.

Hence, discarding the physically irrelevant square root we have the expression for the power in the l–th subchannel

$$P_l = \frac{A_l}{2\Delta} \frac{\lambda_l - 2\Delta}{\lambda_l - \Delta} - \sqrt{\left(\frac{A_l}{2\Delta} \frac{\lambda_l - 2\Delta}{\lambda_l - \Delta}\right)^2 + \frac{A_l}{\Delta} \frac{A_l - W\lambda_l}{\lambda_l - \Delta}} \qquad (3.22)$$

Setting the sum of positive P_l's (3.22) equal to the overall power P yields the equation for the parameter W. It is identical in form to (3.17). After solution of equation (3.17) and determination of W, expressions (3.22) permit power estimation for each subchannel.

Figure 3.3 illustrates the effect of inter-channel crosstalk. Let us introduce crosstalk into the system considered in Sect. 3.2 with the coefficients (3.19). The crosstalk level is determined by the coefficient Δ, which is assumed to be equal to 0.08. Interference noise is taken to be zero. In addition to the crosstalk, thermal noise must be taken into account.

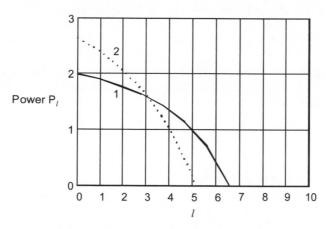

Fig. 3.3. Impact of crosstalk on power distribution over the channels. 1 – no crosstalk ($\Delta = 0$). 2 – crosstalk different from zero ($\Delta \neq 0$)

Curve 1 in Fig. 3.3 faithfully copies Curve 1 of Fig. 3.2 and represents absence of crosstalk. Curve 2 in Fig. 3.3 is plotted with allowance for crosstalk. From comparison of the curves in Fig. 3.3 it transpires that with crosstalk the optimal number of subchannels in the case under consideration decreases from 7 to 6. In the limit case in absence of interference and thermal noise the SISO system becomes an optimal one. In such a case, there is no crosstalk either, and the system capacity grows infinitely.

Thus, while crosstalk hampers the performance of a MIMO transmission system, interference, on the contrary, is beneficial for the efficiency. Difference in crosstalk and interference effects is obvious from the variance expression (3.20). The former appears in it signed negative and the latter signed positive.

We disregard crosstalk in the limit capacity estimations for wireless channels that follow. Crosstalk is detrimental to the efficiency of MIMO systems. In order to secure the capacity of a communication system approaching the limiting one the crosstalk effect should be reduced to a minimum.

3.4 Capacity of Systems with Identical Parallel Channels

The "water-filling" algorithm for optimal power distribution in a communication system has long been known, but has found very limited practical application. Apart from difficult implementation, the added complication is that it provides little improvement in efficiency over the much simpler uniform distribution of power among the channels. Strictly speaking, we cannot use uniform power distribution in limit capacity estimations. With the infinite number of subchannels it will lead to zero capacity.

Uniform power distribution can be used if the number of subchannels with power gains approximately equal and close to the maximum can be established beforehand. Since such estimates are going to be employed in the discussion that follows, let us present now simple capacity formulae.

Compare the capacity C_N of a system having N identical subchannels with the throughput C_1 of a SISO system.

$$C_N = N \log\left(1 + \frac{SNR}{N}\right) \tag{3.23}$$

$$C_1 = \log(1 + SNR) \tag{3.24}$$

In writing (3.23), allowance was made for the N-fold power drop in each of the N subchannels. Figure 3.4 shows the capacity SNR-dependence plots for various N numbers of identical subchannels.

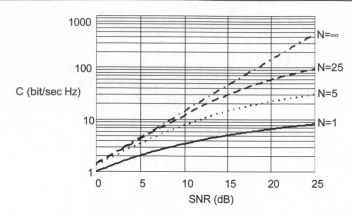

Fig. 3.4. *SNR* dependence of capacity for various numbers of parallel channels

From the above curves and formula (3.23) it is obvious that the capacity of a MIMO system is always bigger than that of a SISO system. Throughput increases as the number of subchannels grows and approaches the limit value equal to

$$\lim_{N \to \infty} C_N = C_\infty = \frac{SNR}{\ln 2} \qquad (3.25)$$

Let us find the number of subchannels N_0, for which the capacity of a MIMO system is close to the limit one (3.25). Assume that it differs from the limit capacity by 20%. Setting (3.23) equal to (3.25) multiplied by 0.8, we obtain the equation for N_0

$$N_0 \log\left(1 + \frac{SNR}{N_0}\right) = \frac{0.8}{\ln 2} SNR$$

Solving the equation for N_0, we obtain the estimation formula for the necessary number of subchannels

$$N_0 = 1.857 \cdot SNR \qquad (3.26)$$

Computations by (3.26) yield the requisite number of subchannels 6, 13 and 186 for *SNR* 5, 10 and 20 dB respectively.

Formula (3.26) yields $N_0 < 2$ with *SNR* < 0 dB. It means that with small *SNR*, employment of MIMO systems makes no sense. Their efficiency grows with an increase of the signal-to-noise ratio.

3.5 Summary

Chapter 3 presents the capacity estimation technique for the multiple input - multiple output communication system. The calculation involves a linear transformation bringing the channel matrix to a diagonal form.

The transformation is followed by working out the Shannon problem for parallel independent channels. The "water filling" algorithm has been presented which will be used subsequently in the book.

The impact of interference and crosstalk as well as of thermal noise on performance of a MIMO system has been discussed. With the overall noise level remaining constant, a rise in the interference was demonstrated to augment the optimal number of the engaged subchannels and the system capacity. A rise in the crosstalk portion of the overall noise leads to quite the opposite effect.

Communication systems with N number of identical subchannels were examined. The required number of subchannels which brings the capacity of a MIMO system to the limiting one has been determined.

4 Analysis of the Two-Dimensional Multipath Channel

In Chapter 4 consideration is being given to two-dimensional multipath wireless channels. In the 2-D wireless channel, all wave sources are located in a single plane (the horizontal, to be more specific.) The distance to the sources is assumed to be big enough to think of the waves produced by them as plane. The vertical-coordinate dependence of the strength of the electric and magnetic field can be neglected.

It is agreed that there is some closed receiving area in the plane. We propose to determine the maximum data transfer rate for a wireless channel with the electromagnetic field picked up in the receiving area.

Chapter 4 is dedicated to solving this problem for the 2-D wireless channel. The same problem for the 3-D channel is worked out in a similar way, but with use of more sophisticated mathematics. We dwell upon the 3-D wireless channel in Chap. 5-8 of the book. Thus, in Chap. 4 we use a simple example to demonstrate solution of problems and the techniques that will be applied to a far more complicated wireless channel in the subsequent chapters of the book.

Mathematical transformations for the plane wave of the 2-D wireless channel are provided in Sect. 4.1. Mathematical treatment of the 3-D channel wave is presented in Chap. 5. Computational techniques for determining the limiting characteristics of a wireless channel, in which the receiver analyzes the electromagnetic field, are available in Sect. 4.2 (for the two-dimensional wireless channel) and in Chap. 6 (for the 3-D wireless channel). The results of numerical estimation of the limiting characteristics are set forth in Sect. 4.3 for the 2-D channel and in Chap. 7 for the three-dimensional channel. Sect. 4.4 and Chap. 8 are dedicated to description of the statistical models of the 2-D and 3-D channel respectively.

Thus, being important in itself, Chap. 4 also holds much methodological significance. It demonstrates by the example of the relatively simple 2-D wireless channel the method which is used later for a more complicated 3-D channel.

4.1 Representation of the Plane Wave Field by a Fourier Series

In a multipath wireless channel the field in the neighborhood of the receiver is a sum of the waves coming from a variety of sources. The latter can be both the primary radiating sources (transmitting antennas) and the secondary ones (scattering objects and reflectors). Let us assume that all the wave sources are in the horizontal plane and at a great distance away from the receiving area. The position of a wave source is characterized by the azimuth coordinate φ_t.

With a distance to the source being great, the wave borne by it can be thought of as plane and the dependence of the observed fields on the vertical coordinate z can be disregarded. In this case, the problem can be looked upon as being two-dimensional, and the components of the electric and magnetic fields can be expressed in terms of two scalar functions [17]. These scalar functions can be viewed as component E_z for the E-wave and component H_z for the H-wave.

The E_z vertical component of the E-wave from a wave source may be written as

$$E_z(\mathbf{r}) = Ee^{-ikR} \tag{4.1}$$

The expression for the H_z field of the H-wave is identical and will not be set out. In (4.1) R denotes the distance between the reference point (r_t, φ_t) and the point of observation (r, φ) – see Fig. 4.1. k stands for the wave number, related to the wavelength λ_w by the expression $k = 2\pi / \lambda_w$. E signifies the complex amplitude of the electric field vertical component at the reference point (r_t, φ_t).

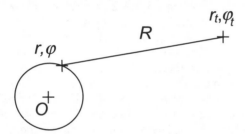

Fig. 4.1. Coordinates of the radiation reference point (r_t, φ_t) and the observation point (r, φ) in the plane

The distance R, as indicated in Fig. 4.1, may be written as

$$R = \sqrt{\left(r_t - r\cos(\varphi - \varphi_t)\right)^2 + \left(r\sin(\varphi - \varphi_t)\right)^2}$$

With the distance to the source being great, $r_t \gg r$, we obtain a simpler expression for R

$$R = r_t - r\cos(\varphi - \varphi_t)$$

Substitution of this expression into (4.1) permits representation of the vertical component of the electric field as

$$E_z(r,\varphi) = E_0 e^{ikr\cos(\varphi - \varphi_t)} \tag{4.2}$$

In (4.2) E_0 denotes the complex amplitude of the electric field vertical component of the wave from the source at the coordinate origin

$$E_0 = Ee^{-ikr_t} \qquad (4.3)$$

For further recasting of (4.2) we make use of the equality [18]

$$e^{ia\cos\psi} = \sum_{n=-\infty}^{\infty} i^n J_n(a)e^{in\psi} \qquad (4.4)$$

Here $J_n(a)$ denotes the Bessel function of the first kind. With negative values of n it is computed with the help of the equation

$$J_{-n}(a) = (-1)^n J_n(a) \qquad (4.5)$$

Expression (4.4) permits writing (4.2) as

$$E_z(r,\varphi) = E_0 \sum_{n=-\infty}^{\infty} i^n J_n(kr)e^{-in\varphi_t} e^{in\varphi} \qquad (4.6)$$

Formula (4.6) demonstrates that observation of the plane wave field along the circumference of a fixed radius r, permits representation of the field in the form of a Fourier series

$$E_z(\varphi) = \frac{1}{2} \sum_{n=-\infty}^{\infty} c_n e^{in\varphi} \qquad (4.7)$$

It is apparent from comparing (4.7) to (4.6) that the c_n complex coefficients of the Fourier series for the plane wave can be calculated by

$$c_n = 2E_0 i^n J_n(kr)e^{-in\varphi_t} \qquad (4.8)$$

For the arbitrary field $E_z(\varphi)$ they can be computed as complex spectral coefficients

$$c_n = \frac{1}{\pi} \int_0^{2\pi} E_z(\varphi)e^{-in\varphi} d\varphi \qquad (4.9)$$

So, the electric field of a plane wave, detected along the circumference, can be written as a Fourier series (4.7). The expansion coefficients are calculated by formula (4.8). We can write absolutely identical expressions for the vertical component of the magnetic field of a plane H-wave.

$$c_n^H = 2H_0 i^n J_n(kr)e^{-in\varphi_t} \qquad (4.10)$$

Here H_0 denotes the complex amplitude of the vertical component of a plane H-wave, φ_t stands for the azimuth direction of its arrival.

Formulae (4.8) and (4.10) permit us to describe the wireless channel as a linear system that converts the input complex amplitudes of the plane waves into the output complex coefficients c_n $(c_n{}^H)$. The c_n $(c_n{}^H)$ coefficients are determined by the vertical components of the electric and magnetic fields.

4.2 Calculation Formulae for Analysis of Two-Dimensional Multipath Wireless Channels

In this section we give formulae for calculating the limit capacity of a multipath wireless channel. Computation of the **H** channel matrix is based on the formulae of Sect. 4.1. After calculating the **H** array, the matrix channel analysis technique described in Sect. 3.2 is applied.

We assume that the field in the neighborhood of the receiver is a sum of plane waves (rays) from a variety of sources. It is further assumed that the sources are located at different angles and the amplitudes of their waves can vary and be specified independently of each other.

In estimation of the limit capacity of a wireless channel we have to believe that an equal amount of the primary source power goes into formation of each individual ray. This assumption is an idealization, but it is due to this idealism that the channel capacity reaches its limit value. Should we operate on the premise that the amount of power expended for ray formation varies, we would have to exclude weak rays from consideration. This will in turn lead to a narrower range of utilized angles, a decrease in the number of involved spatial subchannels, and, finally, to a diminished capacity.

Thus, the number of sources is held to be big enough and their amplitudes are taken to be variable and settable independently of each other. Availability of multiple independent wave sources is one of the prerequisites for building high data-rate multichannel communication systems. With the assumption that this condition is met, we focus on how the size of the receiving area, the layout of the wave sources and other parameters influence the limit characteristics of a wireless communication system.

The linear conversion of the waves of one type in a wireless channel is described by the array equality

$$\mathbf{B} = \mathbf{H} \cdot \mathbf{A} \tag{4.11}$$

Here **A** denotes the M-size input vector of the plane wave complex amplitudes (M specifies the number of sources), **B** stands for the N-size output vector and **H** signifies the $N \times M$ matrix of the wireless channel.

The components of the **B** output vector will be assigned different meanings. They may represent the complex amplitudes of the electric field vertical component E_z and the magnetic field vertical component H_z at the discrete points located within the receiving area or along its perimeter. Then the size N of the **B** vector is equal to the number of discrete points where the electromagnetic fields are detected.

With a circular receiving area the components of the **B** vector will be understood as the complex amplitudes c_n of the spatial harmonics, tackled in Sect. 4.1. The size of the **B** vector then equals the number N of the detected harmonics.

It is apparent that the type of the calculation relationships for the elements of the **H** channel matrix is dependent on the shape of the receiving area, the assumed arrangement of the discrete points in the receiving area and their numeration order. Let us now consider these issues that may be classed with the category of computational techniques.

First of all, notice that the wave sources will be numbered in order of the increasing azimuth angle φ.
$m = 0, 1, 2, \ldots, M\text{-}1$.
We shall examine a 2-D omnidirectional wireless channel and a wireless channel with the wave sources located within some angular sector. For the omnidirectional channel the angles of wave arrival will be determined by the equality

$$\varphi_m = m \frac{2\pi}{M} \tag{4.12a}$$

For the sources located within the angular sector $\Delta\varphi$

$$\varphi_m = -\frac{\Delta\varphi}{2} + m\frac{\Delta\varphi}{M-1} \tag{4.12b}$$

Let us consider now what form the calculation relationships for the H array elements assume during detection of the amplitudes of the harmonic constituents c_n of the vertical field component. The running n has normal numeration
$n = 0, 1, 2, \ldots, N-1$.
The index of the Bessel functions in (4.6-4.10) is symbolized as $n1$. It takes on both positive and negative values. Let us establish the following correspondence between the index $n1$ and the number n

$$n1(n) = \begin{cases} n/2 \ldots\ldots\ldots\ldots for\ldots even\ldots n \\ -(n+1)/2 \ldots for\ldots odd\ldots n \end{cases}$$

Then, as is clear from (4.8), the elements of the **H** matrix array should be calculated by the formula

$$H_{n,m} = ((-1)^n i)^{n1(n)} J_{|n1(n)|}(2\pi r) e^{-in1(n)\varphi_m} \tag{4.13}$$

The multiplication factor 2 common for all the elements has been omitted in writing (4.13). The waves of one type only (the E-type) were taken into account in writing (4.13). There is good reason to believe that an absolutely identical expression can be written for H-waves. This assumption implies that detection of the magnetic field of the H-waves permits conveying information exactly in the same manner as detection of the electric field of the E-waves. The presence of the waves of the second type is allowed for during computations by doubling the number of spatial subchannels with perfectly identical eigenvalues (3.8) as in the case with the waves of one type only.

While registering field vertical components at the discrete point of the receiving area the **H** channel array should be written on the basis of expression (4.2). Then the $H_{n,m}$ elements of the channel matrix can be calculated by the formula

$$H_{n,m} = e^{ikr_n \cos(\varphi_n - \varphi_m)} \tag{4.14}$$

In (4.14) (r_n, φ_n) signify the coordinates of the points where the vertical components of the electric and magnetic fields are determined. The coordinate computation procedure for areas of various shapes is different. Later in this Chapter we examine three types of receiving areas, whose limit capacity may be calculated by means of formula (4.14). These are a circular area, a straight line (a linear lattice) and a rectangular area. The characteristics of the circular area can be calculated in two ways by (4.13) and (4.14). The similarity of the obtained results attests to the accuracy of the computational formulae.

Let us present the formulae for calculating the coordinates of the field detection points for various receiving areas. During field detection at the discrete points located along a circumference (Fig.4.2) the angular coordinates will be computed by

$$\varphi_n = n\frac{2\pi}{N} \tag{4.15}$$

Clearly, the radial coordinate in so doing remains unchanged $r_n = r$.

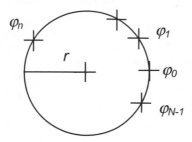

Fig. 4.2. Arrangement of the discrete field pickup points in a circular receiving area

Let us take a look at a disposition of the discrete points on a rectilinear segment (the linear lattice shown in Fig.4.3). In this case the radial coordinate of the detection points is found by the formula

$$r_n = \left| -a + n\frac{2a}{N-1} \right| \tag{4.16}$$

$2a$ denotes the size of the line segment. The angular coordinate equals $\varphi_n = \varphi_0$, if the expression under the modulus sign in (4.16) is positive, otherwise $\varphi_n = = \varphi_0 + \pi$.

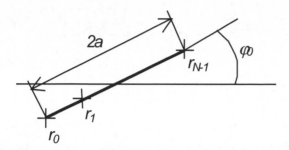

Fig. 4.3. Location of the discrete field detection points along the rectilinear segment

Consider a rectangular receiving area of size a along the x axis and size b along the y axis. The approximate value of the distance Δ_{ap} between the discrete points is also assumed to be provided. We shall compare the system where field detection occurs along the rectangle perimeter (Fig. 4.4a), with the system in which the pickup points are strewn within the confines of the rectangular area (Fig. 4.4b). With the discrete points located along the periphery, their Cartesian coordinates are calculated by:

$$x_n = \begin{cases} -\dfrac{a}{2} + \dfrac{a}{N_x}\left[\dfrac{n}{2}\right] \dots\dots for \dots n < 2(N_x + 1) \\[3mm] (-1)^{n-1}\dfrac{a}{2} \dots\dots\dots for \dots n \geq 2(N_x + 1) \end{cases} \tag{4.17a}$$

$$y_n = \begin{cases} (-1)^{n-1}\dfrac{b}{2} \dots\dots\dots for \dots n < 2(N_x + 1) \\[3mm] -\dfrac{b}{2} + \dfrac{b}{N_y}\left[\dfrac{n - 2N_x}{2}\right] \dots for \dots n \geq 2(N_x + 1) \end{cases} \tag{4.17b}$$

The overall number of the detection points is

$$N = 2(N_x + N_y) \tag{4.17c}$$

The values L_x and L_y are determined by:

$$N_x = \left[\dfrac{a}{\Delta_{ap}}\right], \quad N_y = \left[\dfrac{b}{\Delta_{ap}}\right] \tag{4.17d}$$

In (4.17) the square brackets $[x]$ denote the integer part of x

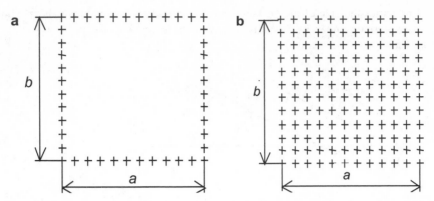

Fig. 4.4. Disposition of the field pickup points in a rectangular receiving area. **a** – pickup points located along the area perimeter, **b** – discrete points scattered within the receiving area

The coordinates of the detection points distributed within the entire rectangular area (Fig. 4.4b) are defined by:

$$x_n = -\frac{a}{2} + \frac{a}{N_x} \operatorname{mod}(n, N_x + 1) \tag{4.18a}$$

$$y_n = -\frac{b}{2} + \frac{b}{N_y}\left[\frac{n}{N_x + 1}\right] \tag{4.18b}$$

The overall number of the detection points is given by

$$N = (N_x + 1)\cdot(N_y + 1) \tag{4.18c}$$

The quantities N_x and N_y are defined by the earlier expressions (4.17d). The function $\operatorname{mod}(n, N)$ signifies the residue of division n by N.

In formulae (4.13 – 4.14) waves of one type only are taken into account. To allow for the waves of the second type during a capacity estimation one should double the number of the eigenvalues of the $\mathbf{H^{+}H}$ ($\mathbf{H \cdot H^{+}}$) matrix arrays as was mentioned earlier in discussion of (4.14).

The formulae of this section and the calculation relationships of Sect. 3.2 permit estimation of the limit capacity of a two-dimensional wireless channel. The numerical results of the calculation are presented in Sect. 4.3.

4.3 Limit Capacity of the Two-Dimensional Multipath Wireless Channel: Analysis Results

In this section we present the results obtained during a numerical analysis of the limit capacity of a multipath wireless channel with detection of the electromagnetic field in the receiving area. Three types of receiving areas are considered: a circle, a rectilinear segment and a rectangle. The numerical analysis results are described in Sect. 4.3.1, 3.3.2 and 4.3.3 respectively.

We analyze an omnidirectional wireless channel with uniform distribution of sources within an azimuth angle, and also examine a sector channel with sources located within some angle sector. We compare the characteristics of systems with thermal and interference noise at the channel output.

We will further look into how various parameters influence the characteristics of the systems. The size of the receiving area, the number of wave sources, the quantity of field pickup points and SN ratios will be varied. The section contains a considerable amount of the plots that illustrate the dependence of the optimal number of spatial subchannels and the capacity per unit bandwidth of a multipath wireless channel on a variety of parameters. Other dependencies of interest to the reader can be derived on the basis of the formulae discussed in Chap. 2 and Sect. 4.2.

4.3.1 Limiting Characteristics of a Wireless System with Analysis of the Electromagnetic Fields along a Circumference

In this subsection we concern ourselves with a system having a circular receiving area. The channel matrix for the system can be calculated both by means of (4.13) and by (4.14). Expression (4.13) permits writing approximate formulae for evaluation of the channel characteristics. The capacity can be approximately estimated by means of formula (3.23), with the number N_{ap} of roughly equipotent subchannels in the system provided by

$$C_{ap} = N_{ap} \log\left(1 + \frac{SNR}{N_{ap}}\right) \tag{4.19}$$

The number of the equipotent spatial subchannels is readily estimated for a small-radius receiving area. With a small-size receiving area one summand becomes predominant in the (4.6) series. With $x \ll 1$ $J_0(x) \approx 1$ and $J_n(x) \ll 1$ for $n \neq 0$. Therefore, the number of spatial subchannels becomes equal to two (one subchannel for each type of the waves, the E- and H-type).

$$N_{ap} = 2 \tag{4.20}$$

Only polarization diversity is efficacious in the 2-D channel with a small-size receiving area. It is pertinent to note that in the 2-D model a receiving area of small size does not imply use of electrically small antennas. The vertical dimension of the antenna may be big. Electrically small antennas lead to a different result when the optimal number of subchannels is in excess of two. The results of analyzing electrically small antennas are expounded in Chap. 7 where 3-D wireless channels are considered.

In estimation of the number of spatial subchannels for a big-radius receiving area it is well to bear in mind that the values of the $J_n(x)$ Bessel function decrease rapidly as the n running number grows with $n >> x$. This property of the Bessel function is utilized in bandwidth estimation of the angle modulation signal. Therefore, in the formally infinite Fourier series (4.6) with $kr >> 1$ we may restrict ourselves to $n = kr$. In such a case the number of spatial subchannels may be estimated by

$$N_{ap} = 2(2kr + 1)$$

(4.21a)

Multiplier 2 in expression (4.21a) accounts for presence of the waves of both types. It is assumed that in addition to the E-waves, H-waves are also available. For the waves of one type only factor 2 must be omitted. The multiplier 2 in the parentheses allows for negative and positive values of the n index of the Bessel function in (4.6). Expression (4.21a) can also be rewritten in terms of the wavelength λ_w.

$$N_{ap} = 2\left(\frac{2l}{\lambda_w} + 1\right)$$

(4.21b)

Here l denotes the circumference that encloses the receiving area. Therefore, an omnidirectional wireless channel contains approximately four times as many spatial subchannels as there are waves (in terms of the wavelength) that fit along the circumference enclosing the receiving area.

Formulae (4.19) and (4.21) must be regarded as being tentative and rough. They do not make any allowance for the number of subchannels being dependent on SNR, the number of the wave sources, and many other factors. The effect of these factors defies all attempts of analytical estimation. This being so, we proceed with considering the results of the numerical analysis.

Initially we will look at an omnidirectional channel with thermal noise at the output. The channel matrix is calculated by formula (4.13). After that, on the strength of this matrix and the formulae of Sect. 3.2, the optimal number of subchannels and the limit capacity of the system are computed. The SNR-dependence plot for the optimal number of subchannels with an electrically small receiving area ($r = 0.04\lambda_w$) is shown in Fig. 4.5a , the SNR-dependence of the limit capacity is demonstrated in Fig. 4.5b. For reference we have included the SNR-dependence of a SISO system in the same Fig.

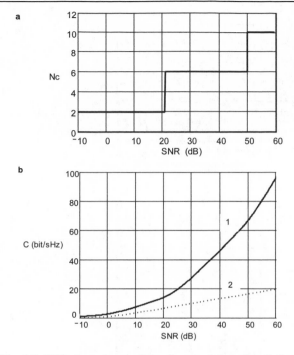

Fig. 4.5. SNR dependence of the characteristics of a wireless channel where **a** presents the SNR-dependence of the optimal number of subchannels and **b** gives the SNR-dependence of the limit capacity per unit bandwidth. Curve 1 stands for the optimal power distribution over the subchannels, and Curve 2 represents the capacity of a SISO system. Analysis of the vertical field components along the circumference of a small-radius circle. The circle radius $r = 0.04\lambda_w$, thermal noise $d^T = 1$, the number of wave sources $M = 130$, the number of analyzed harmonics $N = 91$.

The plot in Fig. 4.5a substantiates the conclusion that with a small-size receiving area the number of subchannels remains two in a wide range of SNR values. The number of the subchannels increases to 6 if the SNR is in excess of approximately 20 dB, and grows further to 10 with the SNR beyond 45 dB. Such a big number of subchannels with a small-size receiving area implies use of superdirective receiving antennas. That is, the main impediment to employment of superdirectivity is a limited signal-to-noise ratio.

Fig. 4.5b gives the SNR dependence of the limit capacity (Curve 1) and the SNR dependence of the capacity of a SISO system (Curve 2). Comparison of the curves lends support to the inference of Sect. 3.4 about poor efficiency of the SISO system with low SNR. The gain in the capacity of the MIMO system grows with a rise in SNR. For example, with SNR = 10 dB the maximum capacity of the MIMO system equals 5.17 (bit/sHz). The capacity of the SISO system is 3.46 (bit/sHz) with SNR = 10 dB.

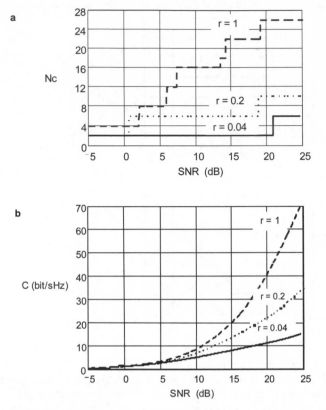

Fig. 4.6. SNR dependence of the characteristics of an omnidirectional wireless channel where **a** presents the SNR dependence of the optimal number of subchannels and **b** gives the SNR dependence of the limit capacity. Analysis of the field vertical components along the circumference. The radius of the circumference is expressed in terms of λ_w fractions, the noise is thermal (thermal noise variance $d^T = 1$), the number of wave sources $M = 130$, the number of harmonics under analysis $N = 91$.

The plots in Fig. 4.6a and 4.6b present the variations of the characteristics in response to changes in the radius of the receiving area. The computations have been performed for three radius values ($r = 0.04\ \lambda_w$, $r = 0.2\ \lambda_w$, $r = \lambda_w$).

It can be seen from Fig. 4.6 that as the radius increases so does the optimal number of subchannels and the maximum capacity of the system. The gain in the throughput is most pronounced with big SNR values. Thus, with the multipath omnidirectional wireless channel it is not so much big spatial dimensions as big SNR that is important for building an efficient system.

During calculation of the characteristics presented in Fig. 4.7 interference noise was assumed to prevail. The level of thermal noise is −10dB of the overall noise.

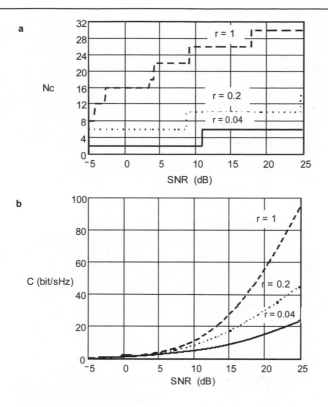

Fig. 4.7. The SNR dependence of the characteristics of the omnidirectional wireless channel where **a** presents the SNR dependence of the optimal number of subchannels and **b** gives the SNR dependence of the limit capacity. Analysis of field vertical components along the circumference. The radius of the circumference is expressed in terms of λ_w fractions, interference noise prevails (thermal noise variance $d^T = 0.1$), the number of wave sources $M = 130$, the number of harmonics under analysis $N = 91$.

A comparison of the plots in Fig. 4.7 with the plots in Fig. 4.6 substantiates the conclusion that with interference noise MIMO systems are more efficient than with thermal noise. For the circumference radius $r = \lambda_w$ with SNR = 20 dB the limit capacity is 55.8 (bit/sHz). As Fig. 4.7 suggests, the system must have 30 spatial subchannels to attain such a capacity.

Figure 4.8 and Fig. 4.9 present the results of the inquiry into a sector channel. It is assumed that the wave sources are evenly distributed within a 60 degree angle sector. During plotting the graphs of Fig. 4.9 it was assumed that interference noise prevails (thermal noise level being −10dB of the overall noise). The results depicted in Fig. 4.8 are representative of the absence of interference noise. The obtained evidence points to the fact that in the sector channel a receiving area of a much bigger size than that in the omnidirectional channel is needed to establish a MIMO system.

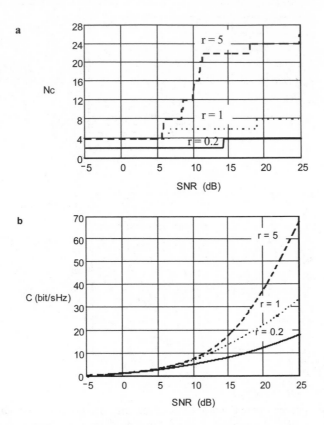

Fig. 4.8. SNR dependence of the characteristics of the sector (60 degrees) wireless channel where **a** presents the SNR dependence of the optimal number of subchannels and **b** gives the SNR dependence of the limit capacity. Analysis of field vertical components along the circumference. The radius of the circumference is expressed in terms of λ_w fractions, there is no interference noise (thermal noise variance $d^T = 1$), the number of wave sources $M = 130$, the number of harmonics under analysis $N = 91$.

When plotting the graphs in Fig. 4.8 and 4.9 the radius values were assumed 5 times the radius values in the plots of the omnidirectional wireless channel in Fig. 4.6 and 4.7.

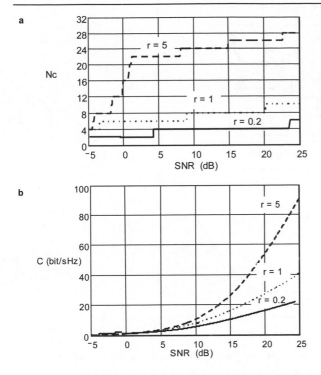

Fig. 4.9. SNR dependence of the characteristics of the sector (60 degrees) wireless channel where **a** presents the SNR dependence of the optimal number of subchannels and **b** gives the SNR dependence of the limit capacity. Analysis of field vertical components along the circumference. The radius of the circumference is expressed in terms of λ_w fractions, interference noise prevails (thermal noise variance $d^T = 0.1$), the number of wave sources $M = 130$, the number of harmonics under analysis $N = 91$.

Figure 4.10 illustrates how the limiting characteristics vary with a change in the number of analyzed harmonics. The wireless channel is assumed to be omnidirectional, thermal noise $d^T = 1$, the radius of the circumference enclosing the receiving area is taken to be equal to λ_w.

Curves 1, 2 and 3 in Fig. 4.10 represent different quantities of the analyzed harmonics: 15, 9 and 5 for each of the two wave types. The maximum possible quantities of spatial subchannels in these cases are 30, 18 and 10. The plots in Fig. 4.10 indicate that if the maximum possible number of spatial subchannels exceeds their optimal number in Fig. 4.6 ($r = 1$), then a limitation of the number of analyzed harmonics does not influence the system characteristics. The initial segments of the plots in Fig. 4.10 and 4.6 coincide. However, when the maximum possible number of spatial subchannels is less than their optimal number, a decrease in the number of analyzed harmonics leads to a smaller number of spatial subchannels and reduces the maximum attainable capacity of the channel.

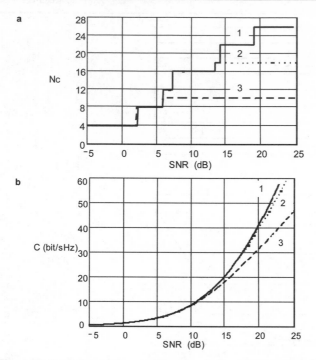

Fig. 4.10. Variation of the omnidirectional channel characteristics with a change in the N number of the analyzed spatial harmonics where **a** presents the SNR dependence of the optimal number of subchannels and **b** gives the SNR dependence of the limit capacity. Analysis of the vertical field components along the circumference. The circle radius equals λ_w, the noise is thermal ($d^T = 1$), the number of wave sources $M = 50$. $1 - 2N = 30$, $2 - 2N = 18$, $3 - 2N = 10$.

Figure 4.11 portrays, the changes in the limiting characteristics following the variations in the number of plane waves. The wireless channel is taken to be omnidirectional, thermal noise $d^T = 1$, the radius of the circumference enclosing the receiving area is taken to be equal to λ_w.

Curves 1, 2 and 3 in Fig. 4.11 reflect different wave source numbers 15, 9 and 5 for horizontal and vertical polarization. The maximum possible number of spatial subchannels in these cases is 30, 18 and 10. Our interest is in comparing the plots in Fig. 4.11 and 4.10. The maximum possible number of spatial subchannels is the same for big SNR values. However in Fig. 4.11, with a decrease in the number of the sources, the values of the characteristics also vary for all SNRs. Still, with a smaller number of harmonics, as is indicated by Fig. 4.10, only some of the values change. This clearly demonstrates that description of the multipath wireless channel in terms of spatial harmonics is preferable to its ray-based description. We will look more closely at the interrelation of these two conceptualizations of the wireless channel in Sect. 4.4.

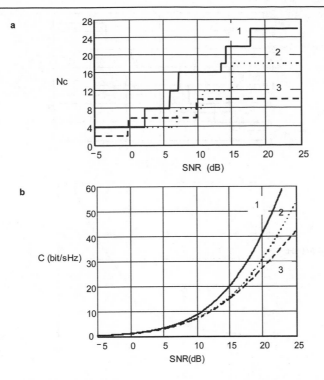

Fig. 4.11. Variations in the characteristics of the omnidirectional wireless channel following the change in the M number of wave sources. $1 - 2M = 30$, $2 - 2M = 18$, $3 - 2M = 10$. Chart **a** presents the SNR dependence of the optimal number of subchannels. Chart **b** gives the SNR dependence of the limit capacity. The overall number of the vertically and horizontally polarized wave sources equals $2M$. Analysis of the vertical field components along the circumference. The circle radius is equal to λ_w, the noise is thermal ($d^T = 1$).

While computing the rest of the channel characteristics in Sect. 4.3, the channel matrix should be calculated by formula (4.14), and not by (4.13). This implies that different components of the electric and magnetic fields are detected at the discrete points of the receiving area. It permits study of the receiving areas of shapes other than circular and analysis of the effect that the change in the number of field pickup points has on the channel characteristics. As before, the formulae of Sect. 3.2 are used for calculating the optimal number of subchannels and the limiting characteristics.

The plots of Fig. 4.12 demonstrate the effect that the quantity of vertical field component detection points has on the system characteristics. The discrete points are located along the circumference (see Fig. 4.2). We consider an omnidirectional wireless channel with 50 sources of vertically polarized waves and 50 sources of horizontally polarized waves. The circle radius equals λ_w, thermal noise ($d^T = 1$).

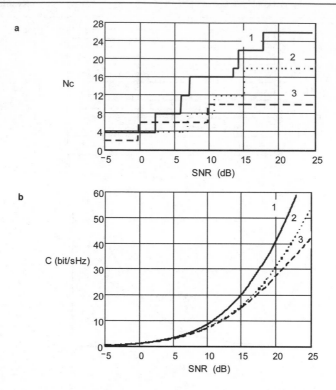

Fig. 4.12. Variations in the characteristics of the omnidirectional wireless channel following the change in the L number of the field registration points. $1 - L = 15$, $2 - L = 9$, $3 - L = 5$. Chart **a** presents the SNR dependence of the optimal number of subchannels. Chart **b** gives the SNR dependence of the limit capacity. The overall number of vertically and horizontally polarized wave sources equals $2M = 100$. Analysis of the vertical field components along the circumference. The circle radius is equal to λ_w, the noise is thermal ($d^T = 1$).

The number of points picking up the vertical components of the electric and magnetic fields is taken to be 15, 9 and 5.

A comparison of the plots in Fig. 4.12 and 4.11 reveals that a decrease in the number of field detection points causes approximately the same change in the characteristics as brought about by a decrease in the number of wave sources.

After the detailed examination of the circular receiving area let us touch upon receiving areas of different shapes.

4.3.2 Limiting Characteristics of the Wireless Channel in Field Analysis along a Rectilinear Segment

In this subsection we assume that the radiation fields are picked up along a rectilinear segment of the length $2a$ (see Fig. 4.3). The channel matrix is calculated by formula (4.14) with the limiting characteristics being estimated by the formulae of Sect. 3.2.

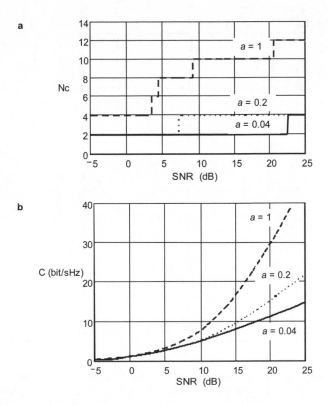

Fig. 4.13. The limiting characteristics of the omnidirectional wireless channel with registration of the vertical field components along a rectilinear segment of length $2a$ where **a** presents the SNR dependence of the optimal number of subchannels and **b** gives the SNR dependence of the limit capacity. The overall number of vertically and horizontally polarized wave sources equals $2M = 260$. The number of field detection points $L = 50$. The noise is thermal ($d^T = 1$).

Figure 4.13a presents the SNR dependence of the optimal number of spatial subchannels and the limit capacity.

A comparison between the above plots and those in Fig. 4.6a shows that in case of the circular receiving area the number of spatial subchannels is greater than in case of the line segment of the length $2a = 2r$. When the limit capacity plots in Fig. 4.13b are compared with those in Fig. 4.6b it is apparent that a circular receiving area provides a bigger capacity.

Figure 4.14 shows the characteristics of a (60 degrees) sector channel. The size of the line segment $2a = 2\lambda_w$.

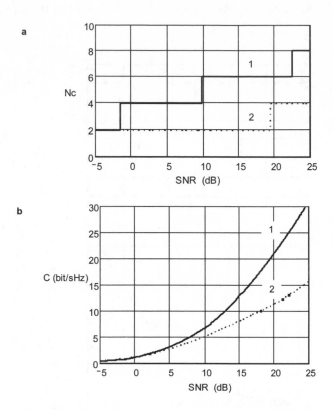

Fig. 4.14. The limiting characteristics of the sector wireless channel with vertical field components being detected along a rectilinear segment of length $2a = 2\lambda_w$ where **a** presents the SNR dependence of the optimal number of subchannels and **b** gives the SNR dependence of the limit capacity. Curve 1 represents the line segment positioned across the direction of wave propagation. Curve 2 corresponds to the line segment positioned along the direction of wave propagation. The overall number of vertically and horizontally polarized wave sources equals $2M = 260$. The number of field detection points $L = 50$. The noise is thermal ($d^T = 1$).

Plots 1 represent crosswise position of the line segment relative to the direction of wave propagation. Plots 2 reflect positioning the line segment along the direction of wave propagation.

A comparison of plots 1 and 2 in Fig. 4.14 demonstrates that longitudinal positioning of the line segment results in materially worse characteristics than with crosswise orientation. This is attributed to the fact that the radius of spatial correlation of the fields is much greater in the longitudinal direction than it is in the transverse direction.

A comparison of the plots in Fig. 4.14 and 4.13 is evidence that the limiting characteristics of the sector wireless channel are much worse than those of the omnidirectional channel.

Figure 4.15 shows the dependence of the limit capacity on the size of the line segment along which field detection occurs. The channel is taken to be omnidirectional, the number of wave sources equals 20, and the number of field detection points along the segment line is 15. The noise is thermal ($d^T = 1$).

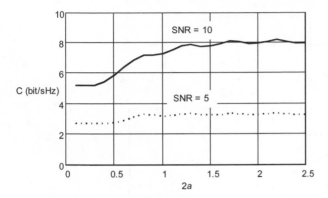

Fig. 4.15. The $2a$ segment length dependence of the channel limiting characteristics

Analysis of the vertical field components along the segment. The segment length is given in fractions of λ_w, the noise is thermal ($d^T = 1$), the number of wave sources $M = 20$, the number of discrete field detection points $N = 15$.

Three regions can be distinguished in the presented plots, corresponding to a small, large and intermediate sized receiving area. With a small-size receiving area the number of spatial subchannels is minimal and equals two, discrimination is possible by polarization and the wave type (E-type and H-type) only.

A large-size receiving area allows a relatively large number of spatial subchannels and the capacity approaches the limit one C_{∞}. With a medium-size area the channel capacity grows together with expansion of the field analysis region. The boundary of the medium-size area is a matter of convention and depends on SNR.

4.3.3 Limiting Characteristics of the Two-Dimensional Multipath Wireless Channel with a Rectangular Receiving Area

In this subsection we present some characteristics of a wireless communication system with a rectangular receiving area. As in 4.3.2, the **H** wireless channel matrix is calculated by formula (4.14).

The plots in Fig. 4.16 apply to the omnidirectional wireless channel with a square receiving area. The square side is taken to be equal to the wavelength λ_w. The vertical components of the electric and magnetic fields are detected. We compare the characteristics of the system where the fields are analyzed along the area perimeter only (Curves 1) with those of the system in which the field analysis is performed both along the perimeter and inside the area (Curves 2).

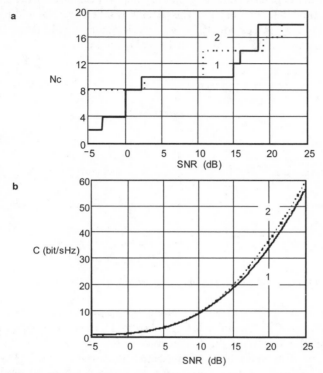

Fig. 4.16. SNR dependence of the omnidirectional channel characteristics where **a** presents the SNR dependence of the optimal number of subchannels and **b** gives the SNR dependence of the limit capacity. Analysis of the field vertical components for the square of the side λ_w. No interference noise (thermal noise variance $d^T = 1$), the number of wave sources $M = 50$. Curve 1 is for the field analysis at the 44 discrete points along the square perimeter. Curve 2 represents the field analysis at the 144 points evenly distributed over the square area

For the purpose of the analysis the side of the square is divided into 11 segments. Thus, it is assumed that the fields are picked up either at the 44 points along the square perimeter as in Fig. 4.4a, or at 144 points of the uniform grid of the square area as in Fig. 4.4b.

A comparison of Curves 1 and 2 in Fig. 4.16b indicates that introduction of additional detection points inside the receiving area does not contribute much to the limit capacity. This evidence strengthens the theoretical deduction about the peripheral points of the receiving area containing the entirety of transmitted information. At the same time this does not mean that field analysis at the internal points of the receiving area is of no value. Field analysis at the internal points can simplify the communication system. The known signals detected at these points can be regarded as reference symbols that decrease the likelihood of error in the system. That is, introduction of spatial elements (additional field pickups) into a wireless communication system can shift the focus from time coding techniques to spatial methods.

The plots in Sect. 4.3 of course do not give an exhaustive account of all cases of interest. The limiting characteristics of wireless channels with receiving areas of other shapes or with different parameters can be calculated with the help of the formulae presented in Chap. 2 or Sect. 4.2.

4.4 Statistical Model of the Multipath Wireless Channel Based on Polar Solution to Maxwell's Equation

In the multipath channel model accepted in literature [1-3] the electromagnetic field in the receiving area is defined as a sum of diversely directed plane waves. The plane wave (ray) is a Cartesian solution to Maxwell's equation. We use a different approach to building a statistical model of the wireless channel in this Section.

The electromagnetic field in the receiving area is regarded as a sum of basis functions providing polar coordinate solutions to the Maxwell equations. The statistical properties of the model are determined by the statistical properties of the random coefficients by which the basis functions are multiplied. Let us turn to discussion of the proposed model and find out how it is related to the commonly accepted ray-oriented model.

It is well known that the polar coordinate solution to Maxwell's equations is a sum of the waves of two types [19]. These are the E- and H-waves. All the components of the electric and magnetic fields can be expressed in terms of the E_z component for the E-wave and the H_z component for the H-wave.

The expressions for a two-dimensional problem, with the fields being independent of the z vertical coordinate, are of the following form. For E-waves:

$$E_\varphi = E_r = 0 \,, \; H_z = 0$$

$$H_\varphi = \frac{i}{k}\sqrt{\frac{\varepsilon}{\mu}}\frac{\partial E_z}{\partial r} \,, \; H_r = -\frac{i}{kr}\sqrt{\frac{\varepsilon}{\mu}}\frac{\partial E_z}{\partial \varphi} \tag{4.22a}$$

For H-waves:

$$E_\varphi = -\frac{i}{k}\sqrt{\frac{\mu}{\varepsilon}}\frac{\partial H_z}{\partial r} \,, \; E_r = \frac{i}{kr}\sqrt{\frac{\mu}{\varepsilon}}\frac{\partial H_z}{\partial \varphi} \tag{4.22b}$$

$$E_z = 0 \,, \; H_\varphi = H_r = 0$$

The E_z and H_z fields appearing in these formulae are determined by the expressions:

$$E_z = \sum_{n=-\infty}^{\infty} \alpha_n i^n J_n(kr) e^{in\varphi} \tag{4.23a}$$

$$H_z = \sum_{n=-\infty}^{\infty} \beta_n i^n J_n(kr) e^{in\varphi} \tag{4.23b}$$

In the statistical model of a wireless channel the coefficients α_n and β_n represent random quantities. Their statistical properties determine the statistical properties of the model.

Formulae (4.23) define the vertical components of the fields. Other cylindrical components are computed by formulae (4.22). Let us write the expressions for all the nonzero field components. For the E-waves they take the form:

$$E_z = \sum_{n=-\infty}^{\infty} \alpha_n i^n J_n(kr) e^{in\varphi} \tag{4.24a}$$

$$H_\varphi = \sqrt{\frac{\varepsilon}{\mu}} \sum_{n=-\infty}^{\infty} \alpha_n i^{n+1} J_n^1(kr) e^{in\varphi} \tag{4.24b}$$

$$H_r = \frac{1}{kr} \sqrt{\frac{\varepsilon}{\mu}} \sum_{n=-\infty}^{\infty} n\alpha_n i^n J_n(kr) e^{in\varphi} \tag{4.24c}$$

In (4.24b) $J_n^1(x)$ denotes a derivative of the Bessel function. For the H-waves the expressions for the nonzero components of a field have the appearance:

$$H_z = \sum_{n=-\infty}^{\infty} \beta_n i^n J_n(kr) e^{in\varphi} \tag{4.25a}$$

$$E_\varphi = \sqrt{\frac{\mu}{\varepsilon}} \sum_{n=-\infty}^{\infty} \beta_n i^{n-1} J_n^1(kr) e^{in\varphi} \tag{4.25b}$$

$$E_r = \frac{1}{kr} \sqrt{\frac{\mu}{\varepsilon}} \sum_{n=-\infty}^{\infty} n\beta_n i^{n-2} J_n(kr) e^{in\varphi} \tag{4.25c}$$

By means of the angle (the φ-th) and radial components we can determine field projections to the horizontal axes of Cartesian coordinates:

$$E_x = E_r \cos\varphi - E_\varphi \sin\varphi \tag{4.26a}$$

$$E_y = E_r \sin\varphi + E_\varphi \cos\varphi \tag{4.26b}$$

In a similar manner expressions for the Cartesian components of the magnetic field can be written.

To determine the statistical properties of the coefficients α_n and β_n let us look into their relationship with the amplitudes of plane waves. From this point on, we consider the coefficients α_n only, since the statistical properties of the coefficients β_n are identical with those of the coefficients α_n.

The expression for the coefficients α_n in terms of the angle of incidence φ_t of the plane wave and its complex amplitude E_0 is readily obtained from a comparison of expressions (4.6) and (4.24a).

$$\alpha_n = e^{-in\varphi_t} E_0 \qquad (4.27)$$

If the number of plane wave sources is M, then the coefficient α_n is determined by a sum of the summands of the form (4.27) for various φ_m angles of arrival of the plane waves

$$\alpha_n = \sum_{m=0}^{M-1} e^{-in\varphi_m} E_m \qquad (4.28)$$

In (4.28) E_m denotes the complex amplitude of the vertical component of the electric field of the m–th source at the coordinate origin (with $r = 0$).

Assuming, that the angles φ_m form a uniform grid in the range from 0 to 2π

$$\varphi_m = m\frac{2\pi}{M}$$

then (4.28) represents a direct discrete Fourier transform

$$\alpha_n = \sum_{m=0}^{M-1} E_m e^{-inm\frac{2\pi}{M}} \qquad (4.29)$$

Consequently, the other way around, the E_m complex amplitudes of the plane waves can be defined by the coefficients α_n through an inverse discrete Fourier transform

$$E_m = \frac{1}{M} \sum_{n=0}^{M-1} \alpha_n e^{inm\frac{2\pi}{M}} \qquad (4.30)$$

The interrelationship (4.29 and 4.30) of the coefficients α_n (β_n) and the amplitudes of the plane wave, described also in [17], permits determining the statistical properties of the coefficients. In the wake of [20], we will examine three distribution cases for the plane wave sources. Case 1 is a homogeneous distribution in the angle φ. Case 2 represents an inhomogeneous distribution. Case 3 differs from Case 2 by presence of a deterministic plane wave.

Case 1. Homogeneous wave distribution in the angle φ

It is assumed that the complex amplitudes of the plane waves are independent random quantities with zero mean. The term *'normal complex random quantity'* means that such a quantity can be written as

$$E = E_R + iE_I$$

Here E_R and E_I denote independent real normal random quantities of the variance $D/2$. In Case 1 with the homogeneous angle distribution, the wave amplitude variances of all the M sources are taken to be the same and equal to D.

Determine the statistical properties of the coefficients α_n. With the above assumptions the mean value of each coefficient is zero.

$$\left\langle \alpha_n \right\rangle = \sum_{m=0}^{M-1} e^{-in\varphi_m} \left\langle E_m \right\rangle = 0 \tag{4.31}$$

The covariance of the coefficients α_{n1} and α_{n2} is calculated based on the expression

$$\left\langle \alpha_{n2} \alpha_{n1}^{*} \right\rangle = \sum_{m1=0}^{M-1} \sum_{m2=0}^{M-1} e^{-i(n2\varphi_{m2} - n1\varphi_{m1})} \left\langle E_{m2} E_{m1}^{*} \right\rangle \tag{4.32}$$

Taking into account the premise of mutual independence of the sources

$$\left\langle E_{m2} E_{m1}^{*} \right\rangle = 0 \;\; \text{with } m1 \neq m2, \;\; \left\langle \left| E_m \right|^2 \right\rangle = D \tag{4.33}$$

we write:

$$\left\langle \alpha_{n2} \alpha_{n1}^{*} \right\rangle = 0 \;\; \text{with } n1 \neq n2 \tag{4.34a}$$

$$\left\langle \left| \alpha_n \right|^2 \right\rangle = MD \tag{4.34b}$$

Thus, the coefficients α_n are independent normal random quantities with zero mean and identical variances. Their variances equal the sum of variances of the plane waves from all the sources. To put it differently, the variance of every coefficient α_n equals the variance of the electric field vertical component. The same deduction is true for the β_n coefficients, which are independent normal random complex quantities as well. The variance of each β_n coefficient equals that of the magnetic field vertical component.

Case 2. Inhomogeneous distribution of wave sources

In Case 2 we assume as before, that the complex amplitudes of the wave sources are independent random quantities with zero mean values. However, in contrast to Case 1, the distribution of sources within the angle φ is taken to be inhomogeneous.

$$\left\langle \left| E_m \right|^2 \right\rangle = D_m \tag{4.35}$$

The mean values of the coefficients α_n remain equal to zero. Derivation of this statement is the same as derivation of (4.31) in Case 1. The covariance expression (4.32) takes the form

$$\left\langle \alpha_{n2} \alpha_{n1}^* \right\rangle = \sum_{m=0}^{M-1} e^{i(n1-n2)\varphi_m} D_m \tag{4.36}$$

Expression (4.36) demonstrates that the covariance coefficient value is dependent on the difference of $n1$ and $n2$. Denoting $n = n2 - n1$ we can write the discrete covariance function as

$$R(n) = \sum_{m=0}^{M-1} e^{-in\varphi_m} D_m \tag{4.37}$$

Assuming that the φ_m angles form a uniform grid in the range from 0 to 2π

$$\varphi_m = m\frac{2\pi}{M}$$

we rewrite (4.37) as

$$R(n) = \sum_{m=0}^{M-1} D_m e^{-inm\frac{2\pi}{M}} \tag{4.38}$$

Relationship (4.38) is a direct discrete Fourier transform. Thus, the covariance function $R(n)$ is determined by a direct discrete transform of the discrete angle spectrum of the plane waves' power.

Passing on to continuous distribution $D(\varphi)$ with $M \to \infty$, we can write

$$R(n) = \frac{1}{2\pi} \int_0^{2\pi} D(\varphi) e^{-in\varphi} d\varphi \tag{4.39}$$

Expressions (4.38) and (4.39) permit writing the formulae for the angular wave power spectrum in terms of the known covariance function $R(n)$. This is an inverse discrete Fourier transform for a discrete spectrum

$$D_m = \frac{1}{M} \sum_{m=0}^{M-1} R(n) e^{inm\frac{2\pi}{M}} \tag{4.40}$$

With a continuous angular spectrum it is the Fourier series

$$D(\varphi) = \frac{1}{2} \sum_{n=-\infty}^{\infty} R(n) e^{in\varphi}$$ (4.41)

The Fourier transform enjoys wide use in radio engineering. For its calculation a fast Fourier transform algorithm can be employed. The covariance function of the α_n coefficients being linked to the waves' angle spectrum through the Fourier transform is one of the advantages of the proposed model.

It is easy to verify that Case 1 is a specific representation of Case 2, when the angle spectrum is constant. In this case (4.38) yields (4.34) that is uncorrelated coefficients α_n of identical variance.

Case 3. Inhomogeneous distribution of random wave sources in presence of a deterministic plane wave

Case 3 differs from Case 2 by presence of the plane wave of a fixed amplitude E_f with the given incidence angle φ_f. In this case the deterministic summands α^f_n, defined by the formula (4.28) are added to the random coefficients α_n, determined for Case 2.

$$\alpha^f_n = e^{-in\varphi_f} E_f$$ (4.42)

In all three cases the computational relationships for the coefficients β_n are identical to those presented for the coefficients α_n. To be more specific, they are obtained by substituting H for the electric field E in the expressions for α_n.

Therefore, the modeling procedure for a multipath wireless channel can be as follows:
- Generation of coefficients α_n and β_n with the statistical properties defined above

For the omnidirectional wireless channel all the coefficients are independent normal random quantities with zero mean. The variances of all the coefficients α_n are the same and equal to the variance of the E_z electric field vertical component, the variances of the coefficients β_n are also identical and equal to the variance of the magnetic field vertical component H_z.
- Calculation of the fields at the spatial points of interest

For calculation of the fields by the α_n and β_n coefficients, formulae (4.24-4.25) are used.
- Determination of the angle spectrum.

When required, the angle spectrum corresponding to the generated coefficients α_n and β_n can be determined. Use the inverse Fourier transform (4.30).
- Estimation of the signal at the antenna output.

The field model should be augmented by the model of the antenna system, which permits determination of the output signals by the known electromagnetic field. Given the modeling results presented below it is assumed that the antenna elements are field pickups. That is to say, various field components are detected at various points of space.

It is noteworthy that in the proposed model of the electromagnetic field no Doppler shift generation is necessary that would account for motion of the receiving antenna. This shift will appear automatically during signal estimation for the output of the receiving antenna in motion. A field varying in space will result in a signal at the antenna output varying in time. The spectrum of such a signal will include the Doppler frequency shift.

To conclude Sect. 4.4 we present some results obtained by means of the proposed multipath wireless channel model. Fig. 4.17 and 4.18 show the plots for the varying intensity of the electric field vertical component following the variations of a spatial coordinate. The E_z field was computed by formula (4.24a). The maximum running n in the formula was taken to be 100, the Bessel functions were added up through the 100-th order. That is, 201 normal random coefficients α_n were generated during modeling.

Under consideration was the intensity variation of the electric field vertical component in two sensors spaced along the y axis. In Fig. 4.17 the separation is $0.1\lambda_w$, while in Fig. 4.18 the spacing equals $2\lambda_w$. The sensors move towards the x axis.

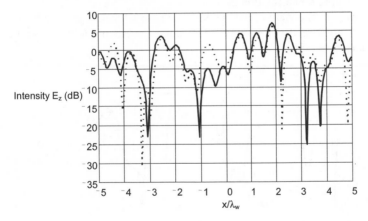

Fig. 4.17. Typical intensity variation of the vertical component E_z of the multipath field with the $0.1\lambda_w$ separation between the sensors

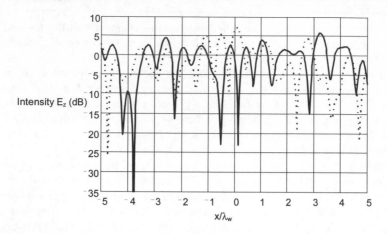

Fig. 4.18. Typical intensity variation of the vertical component E_z of the multipath field with the $2\lambda_w$ spacing between the sensors

Inspection of the presented plots discloses that they have a typical appearance obtainable with the help of the conventional ray-based model [20]. With the $0.1\lambda_w$ spacing in Fig. 4.17 a strong correlation of the plotted curves is noticeable. In the case of the $2\lambda_w$ spacing no correlation is seen.

Figure 4.19 presents the x-coordinate dependence plots for intensity variation of the electric field horizontal components. The curves were plotted by formulae (4.25b, 4.25c and 4.26).

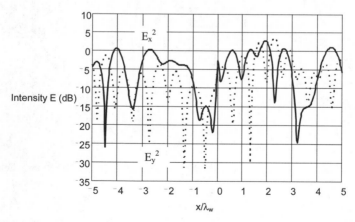

Fig. 4.19. Typical intensity variations of the E_x and E_y components of a multipath field

A comparison of the plots in Fig. 4.19 with the c plots of Fig. 4.17 or 4.18 reveals a certain decline in the maximum intensity of the field horizontal components in contrast to the vertical component. No noticeable correlation of the horizontal field components is observed, though, while plotting the curves of Fig. 4.19 it was assumed that the pickups for the E_x and E_y coincide in space.

A typical angle spectrum is depicted in Fig. 4.20: the azimuth-angle dependence of the E- and H-waves. The presented dependence was calculated by the inverse discrete Fourier transform (4.30). For the purpose of an illustrative rendition, the plot portrays variation of the azimuth angle from 0 through 180^0 rather than through 360^0.

The plots in Fig. 4.20 are evidence that an omnidirectional channel was subject to consideration. The wave amplitudes vary randomly with the changes in the arrival angle.

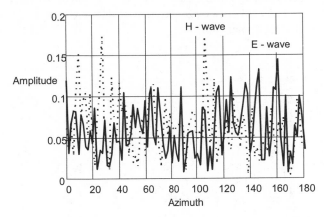

Fig. 4.20. The angle spectrum of E- and H-waves for the omnidirectional wireless channel

Figure 4.21 presents a Doppler spectrum of the electric field vertical component. While calculating it, the E_z component sensor was taken to be in uniform motion along the $-10\lambda_w$, $+10\lambda_w$ segment of the x-axis. The sensor readings are taken at 1024 points of the segment. We further calculated a discrete spectrum of the recorded values. The plot of the spectrum absolute values is shown in Fig. 4.21.

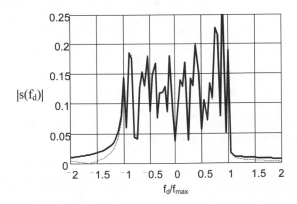

Fig. 4.21. The Doppler spectrum of the E_z field vertical component

In Fig. 4.21 $f_{max} = v/\lambda$ denotes the maximum Doppler shift, and v signifies the velocity of the sensor motion. It is evident from the plot, that the signal spectrum is actually concentrated within the frequency range $-f_{max} < f < f_{max}$. If the frequency is in excess of the maximum Doppler frequency, the spectral amplitudes diminish, approaching zero. However, the spectrum is not identically equal to zero for $|f| > f_{max}$. This is determined by the finite length of the segment and in particular by the fact that the field values in the starting and finishing points of the segment differ ($E_{z0} \neq E_{z1023}$). With the difference smoothed out, the rate of the spectrum's vanishing can be augmented as represented in Fig. 4.21 by the dotted curve.

The spectrum of the E_x and E_y field horizontal components presented in Fig. 4.22 was calculated in a similar manner.

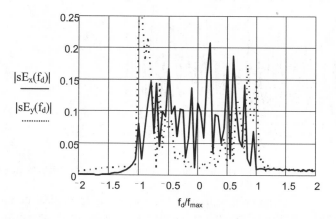

Fig. 4.22. The Doppler spectrum of the E_x, E_y horizontal components of the electric field

It is common knowledge that the nature of the Doppler spectrum of the E_x component differs significantly from that of the E_y component [1]. In the E_x the components close to the zero frequency are predominant while the components approaching the f_{max} are scarce. In the physical perspective it is accounted for by the sensor of the E_x component being positioned along the x-axis with the zero of its pattern directed along the x-axis. Due to this it does not pick up the signals with the maximum Doppler shift. Conversely, in the E_y spectrum the components close to the f_{max}, prevail and the components with the near-zero frequencies are scarce. Inspection of the plots in Fig. 4.22 attests to such a distinction between the spectra.

Thus, the expressions presented in this section permit assessment of various aspects of building a statistical model of the multipath channel.

4.5 Summary

The studies into the two-dimensional multipath wireless channel made in Chap. 4 yielded the following results:

1. The limit capacity formulae and expressions for the optimal number of sub-channels in a multipath wireless channel have been derived. The plots of the *SNR* dependence of the limiting characteristics for 2-D areas of various shapes have been presented.

2. The discussed analytical relationships and plots demonstrate that with a small-size receiving area the optimal number of spatial subchannels equals 2 regardless of the area shape. Only polarization diversity of the subchannels is possible with the limit capacity being that of a system with two spatial subchannels. As the size of the receiving area increases, so does the optimal number of spatial subchannels with the capacity approaching the limit value $C_\infty = SNR/\ln 2$.

3. Examination of the presented plots indicates that in order to attain a material capacity gain it is necessary to proportionally enhance the following parameters:
   − The number of independent wave sources,
   − The size of the receiving area,
   − The signal-to-noise ratio (SNR).

4. A non-ray statistical model of the multipath wireless channel has been proposed. In the proposed model the electromagnetic field in the limited-size receiving area is represented by a sum of polar coordinate solutions to Maxwell's equations with random coefficients. It has been demonstrated that the random coefficients are complex normal random quantities. Their covariance function is a Fourier transform of the angle power spectrum of the wave sources. Specifically, in case of the omnidirectional wireless channel, the random coefficients are independent random quantities with identical variances.

5 Body of Mathematics for Analysis of Three-Dimensional Multipath Wireless Channels: Spherical Harmonics

The commonly accepted visualization of the electromagnetic field in the receiving area consists in thinking of it as a sum of plane waves (rays) produced by various sources. However, such conceptualization is convenient for small quantities of rays only. In case of a multipath channel with a limited-size receiving area, there is good reason to represent the field as an integration of spherical harmonic fields. The following reasoning substantiates the advantage of such conceptualization. First, spherical harmonics allow a vector-to-scalar transition in field description (i.e. a change-over from the strength vectors E and H to electric and magnetic potentials). Second, spherical harmonics make an orthogonal system of functions on the sphere. The coefficients of the electromagnetic field expansion into spherical harmonics allow us to represent a continuous field in terms of a discrete set of complex numbers.

In this chapter we provide some background information on spherical harmonics, and derive spherical harmonics expansion formulae for the plane wave. To put it otherwise, we describe the transition from a continuous ray-oriented concept of the wireless channel to its discrete visualization. The resulting discrete concept of the channel is used in estimating the limit capacity of the 3D wireless channel, and building its statistical model in the subsequent chapters of the book.

Section 5.1 contains the expressions for the electric and magnetic field of spherical harmonics. Section 5.2 provides reference material on the spherical Bessel and associated Legendre functions employed in writing solutions to Maxwell's equations in spherical coordinates. In Sect. 5.3 we exploit the properties of the presented functions in calculating the radiant power of spherical harmonics. Section 5.4 is devoted to writing the expressions for numerical evaluation of the spherical harmonics expansion coefficients for an arbitrary electromagnetic field. Section 5.5 gives a simple example of practical use of the expressions presented in Sect. 5.4 and demonstrates computation of the coefficients for expansion of a dipole radiation field into spherical harmonics.

In Sect. 5.6 we tackle a plane electromagnetic wave propagating along the z axis, and demonstrate an expansion of its field into spherical harmonics. Section 5.7 provides the re-expression formulae for spherical harmonics accounting for rotation of the coordinate system. Finally, in Sect. 5.8 we concern ourselves with determining the coefficients for expansion of the field of an arbitrary plane wave into spherical harmonics.

5.1   Expressing Spherical Three-Dimensional Electromagnetic Fields in Terms of Potentials

We make use of a known fact from electrodynamics that, in a homogeneous medium free from sources, an arbitrary electromagnetic field can be expressed in terms of two scalar coordinate functions. These functions are referred to as the electric (U) and magnetic (V) potentials. For writing the potentials let us use the spherical coordinates depicted in Fig. 5.1.

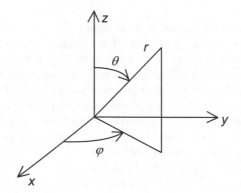

Fig. 5.1. Spherical coordinate system

The U and V spherical potentials may be represented as a total of the summands of the following form [19]

$$U,V = \sqrt{kr}J_{n+0.5}(kr)P_n^{(m)}(\cos\theta)e^{im\varphi} \qquad (5.1)$$

In (5.1) r,θ,φ are the spherical coordinates of a point in Fig. 5.1; n takes on the values $0,1,2,...$ and $m = -n, -(n-1),...,n-1, n$; $k = \omega(\varepsilon\mu)^{0.5}$ denotes the wave number; ε, and μ stand for the dielectric permittivity and magnetic permeability of the medium (vacuum).

$$\varepsilon = \varepsilon_0 = \frac{1}{36\pi}10^{-9}, \quad \mu = \mu_0 = 4\pi10^{-7}$$

ω signifies the angular frequency of the harmonic sources producing a field. $J_{n+0.5}(kr)$ is the Bessel function of the first kind of a half-integer index number $n+0.5$. $P_n^{(m)}(\cos\theta)$ is the associated Legendre function. $P_n^{(0)}(\cos\theta)$ stands for the Legendre polynomial, commonly designated in literature as $P_n(\cos\theta)$. The coordinate-independent normalizing factor is not specified in (5.1).

In what follows, while writing the potentials (5.1) we will use the spherical Bessel functions $j_n(x)$ in lieu of the half-integer index Bessel function $J_{n+0.5}(x)$. Their interrelationship is expressed by the equation presented in [21]

$$j_n(x) = \sqrt{\frac{\pi}{2x}} J_{n+0.5}(x)$$

Therefore, the electric and magnetic potentials can be written as:

$$U(r,\theta,\varphi) = \frac{r}{k} \sum_{n=0}^{\infty} \sum_{m=-n}^{n} \alpha_{n,m} j_n(kr) P_n^{(m)}(\cos\theta) e^{im\varphi} \qquad (5.2)$$

$$V(r,\theta,\varphi) = \frac{r}{k} \sum_{n=0}^{\infty} \sum_{m=-n}^{n} \beta_{n,m} j_n(kr) P_n^{(m)}(\cos\theta) e^{im\varphi} \qquad (5.3)$$

It is worth noting that there are four kinds of spherical Bessel functions. These are the spherical Bessel function of the first kind $j_n(x)$, of the second kind $y_n(x)$ and the spherical Bessel function of the third and fourth kind. The last two, as a rule, are referred to as the spherical Hankel functions of the first $h_n^{(1)}(x)$ and second $h_n^{(2)}(x)$ kind. The $j_n(x)$ function values are confined to the neighborhood of the co-ordinate origin. For this reason, we use expressions (5.2) and (5.3) for writing the fields in a limited area. In order to write a radiation field expression for an unlimited domain, the $h_n^{(2)}(x)$ function should be used instead of function $j_n(x)$. In other words, while considering a field in an unlimited domain the U and V potentials should be written in terms of (5.2) and (5.3), substituting $h_n^{(2)}(x)$ for $j_n(x)$ in them.

All the components of the electric and magnetic fields can be expressed in terms of potentials (5.2) and (5.3) [19]. The E-wave fields are calculated in terms of the electric potential by:

$$E_r = \frac{\partial^2 U}{\partial r^2} + k^2 U \qquad (5.4a)$$

$$H_r = 0 \qquad (5.4b)$$

$$E_\theta = \frac{1}{r} \frac{\partial^2 U}{\partial r \partial \theta} \qquad (5.4c)$$

$$H_\theta = \frac{ik}{r\sin\theta} \sqrt{\frac{\varepsilon}{\mu}} \frac{\partial U}{\partial \varphi} \qquad (5.4d)$$

$$E_\varphi = \frac{1}{r\sin\theta} \frac{\partial^2 U}{\partial r \partial \varphi} \qquad (5.4e)$$

$$H_\varphi = -\frac{ik}{r} \sqrt{\frac{\varepsilon}{\mu}} \frac{\partial U}{\partial \theta} \qquad (5.4f)$$

Substitution of (5.2) into (5.4) gives the field expressions:

$$E_r = \sum_{n=0}^{\infty} \sum_{m=-n}^{n} \alpha_{n,m} \left(\frac{d^2\left(kr j_n(kr)\right)}{d\left(kr\right)^2} + kr j_n(kr) \right) P_n^{(m)}(\cos\theta) e^{im\varphi} \qquad (5.5a)$$

$$E_\theta = \frac{1}{kr} \sum_{n=0}^{\infty} \sum_{m=-n}^{n} \alpha_{n,m} \frac{d\left(kr j_n(kr)\right)}{d(kr)} \frac{d\left(P_n^{(m)}(\cos\theta)\right)}{d\theta} e^{im\varphi} \qquad (5.5b)$$

$$E_\varphi = \frac{i}{kr\sin\theta} \sum_{n=0}^{\infty} \sum_{m=-n}^{n} \alpha_{n,m} m \frac{d\left(kr j_n(kr)\right)}{d(kr)} P_n^{(m)}(\cos\theta) e^{im\varphi} \qquad (5.5c)$$

$$H_r = 0 \qquad (5.5d)$$

$$H_\theta = -\frac{1}{\sin\theta} \sqrt{\frac{\varepsilon}{\mu}} \sum_{n=0}^{\infty} \sum_{m=-n}^{n} \alpha_{n,m} m j_n(kr) P_n^{(m)}(\cos\theta) e^{im\varphi} \qquad (5.5e)$$

$$H_\varphi = -i \sqrt{\frac{\varepsilon}{\mu}} \sum_{n=0}^{\infty} \sum_{m=-n}^{n} \alpha_{n,m} j_n(kr) \frac{d\left(P_n^{(m)}(\cos\theta)\right)}{d\theta} e^{im\varphi} \qquad (5.5f)$$

The link between the field components and the V magnetic potential for H-waves is expressed as in [19]:

$$E_r = 0 \qquad (5.6a)$$

$$H_r = \frac{\partial^2 V}{\partial r^2} + k^2 V \qquad (5.6b)$$

$$E_\theta = -\frac{ik}{r\sin\theta} \sqrt{\frac{\mu}{\varepsilon}} \frac{\partial V}{\partial \varphi} \qquad (5.6c)$$

$$H_\theta = -\frac{1}{r} \frac{\partial^2 V}{\partial r \partial \theta} \qquad (5.6d)$$

$$E_\varphi = \frac{ik}{r} \sqrt{\frac{\mu}{\varepsilon}} \frac{\partial V}{\partial \theta} \qquad (5.6e)$$

$$H_\varphi = \frac{1}{r\sin\theta} \frac{\partial^2 V}{\partial r \partial \varphi} \qquad (5.6f)$$

Substitution of (5.3) into (5.6) gives the expression for the H-wave fields:

$$E_r = 0 \qquad (5.7a)$$

$$E_\theta = \frac{1}{\sin\theta} \sqrt{\frac{\mu}{\varepsilon}} \sum_{n=0}^{\infty} \sum_{m=-n}^{n} \beta_{n,m} m j_n(kr) P_n^{(m)}(\cos\theta) e^{im\varphi} \qquad (5.7b)$$

$$E_\varphi = i\sqrt{\frac{\mu}{\varepsilon}} \sum_{n=0}^{\infty} \sum_{m=-n}^{n} \beta_{n,m} j_n(kr) \frac{d\left(P_n^{(m)}(\cos\theta)\right)}{d\theta} e^{im\varphi} \qquad (5.7c)$$

$$H_r = \sum_{n=0}^{\infty} \sum_{m=-n}^{n} \beta_{n,m} \left(\frac{d^2\left(krj_n(kr)\right)}{d(kr)^2} + krj_n(kr) \right) P_n^{(m)}(\cos\theta) e^{im\varphi} \qquad (5.7d)$$

$$H_\theta = \frac{1}{kr} \sum_{n=0}^{\infty} \sum_{m=-n}^{n} \beta_{n,m} \frac{d\left(krj_n(kr)\right)}{d(kr)} \frac{d\left(P_n^{(m)}(\cos\theta)\right)}{d\theta} e^{im\varphi} \qquad (5.7e)$$

$$H_\varphi = \frac{i}{kr\sin\theta} \sum_{n=0}^{\infty} \sum_{m=-n}^{n} \beta_{n,m} m \frac{d\left(krj_n(kr)\right)}{d(kr)} P_n^{(m)}(\cos\theta) e^{im\varphi} \qquad (5.7f)$$

While dealing with an actual problem, in the radiation field expression the
spherical Hankel function of the second kind $h_n^{(2)}(kr)$ should be substituted for the
spherical Bessel function of the first kind $j_n(kr)$ in formulae (5.5) and (5.7).

5.2   Some of the Properties of Spherical Bessel Functions and Associated Legendre Functions

In this section we present some information about the spherical Bessel functions
$j_n(x)$ and $h_n^{(2)}(x)$, as well as about the associated Legendre functions $P_n^{(m)}(\cos\theta)$,
used in representation of potentials. The information has been borrowed from [19,
21 and 22].

For numerical evaluation of the spherical Hankel function $h_n^{(2)}(x)$ it is possible
to use the following expression

$$h_n^{(2)}(x) = i^{n+1} \frac{e^{-ix}}{x} \sum_{l=0}^{n} \frac{(n+l)!}{l!(n-l)!} (2ix)^{-l} \qquad (5.8)$$

The spherical Bessel function $j_n(x)$ expressed in terms of the spherical Hankel
function $h_n^{(2)}(x)$ is

$$j_n(x) = \mathrm{Re}\left(h_n^{(2)}(x)\right) \qquad (5.9)$$

Re denotes taking the real part of a complex number. The graphs of the functions $j_3(x)$ and $j_5(x)$, plotted with formulae (5.8, 5.9), are shown in Fig. 5.2.

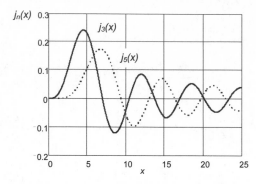

Fig. 5.2. Plots of the spherical Bessel functions

Let us write the explicit expressions for the Bessel spherical functions $j_n(x)$ and $h_n^{(2)}(x)$ for $n \leq 1$.

$$h_0^{(2)}(x) = i\frac{e^{-ix}}{x} \tag{5.10a}$$

$$h_1^{(2)}(x) = -\frac{e^{-ix}}{x}\left(1 + \frac{1}{ix}\right) \tag{5.10b}$$

$$j_0(x) = \frac{\sin x}{x} \tag{5.10c}$$

$$j_1(x) = -\frac{\cos x}{x} + \frac{\sin x}{x^2} \tag{5.10d}$$

Formula (5.8) demonstrates that for the $h_n^{(2)}(x)$ function the following asymptotic expression is true

$$h_n^{(2)}(x) = i^{n+1}\frac{e^{-ix}}{x} \tag{5.11}$$

Although (5.8) and (5.9) look attractively concise, using them would result in unacceptably large computational errors even if the calculations are done on modern computers.

This is why we present a stodgier expression of the spherical Bessel function $j_n(x)$ as a series in terms of powers of x.

$$j_n(x) = \frac{x^n}{1 \cdot 3 \cdot 5 \cdot \ldots \cdot (2n+1)}\left(1 - \frac{\frac{x^2}{2}}{1!(2n+3)} + \frac{\left(\frac{x^2}{2}\right)^2}{2!(2n+3)(2n+5)} \ldots\right) \tag{5.12}$$

For computing the derivative of the spherical Bessel function it is possible to use the following recurrence formula

$$\frac{dj_n(x)}{dx} = j_{n-1}(x) - \frac{n+1}{x} j_n(x), \quad \text{for } n \neq 0 \tag{5.13a}$$

$$\frac{dj_0(x)}{dx} = -j_1(x) \tag{5.13b}$$

Let us turn now to considering the associated Legendre functions $P_n^{(m)}(\cos\theta)$. We use the expression of [22] for numerical computations

$$P_n^{(m)}(\cos\theta) = \left(\frac{1+\cos\theta}{1-\cos\theta}\right)^{\frac{m}{2}} \sum_{l=\max(m,0)}^{n} \frac{(-1)^l (n+l)!}{(n-l)!(l-m)!l!} \left(\frac{1-\cos\theta}{2}\right)^l \tag{5.14}$$

The expression (5.14) given in [22], differs from the formulae presented in [19 and 21] by the factor $(-1)^m$. Presence of the $(-1)^m$ multiplier introduced by Gibson for continuity of the Legendre functions facilitates writing certain calculation formulae. Notice that we should take $P_n^{(m)}(\cos\theta) = 0$ for $|m| > n$.

As an example, we refer to Fig. 5.3 depicting the plot of the $P_{20}^{(5)}(\cos\theta)$ function calculated by formula (5.14).

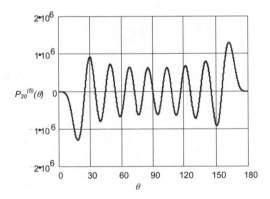

Fig. 5.3. Plot of the associated Legendre function $P_{20}^{(5)}(\cos\theta)$

Consider the explicit expressions of the functions $P_n^{(m)}(\cos\theta)$ for small n and m.

$$P_0^{(0)}(\cos\theta) = 1 \tag{5.15a}$$

$$P_1^{(0)}(\cos\theta) = \cos\theta \tag{5.15b}$$

$$P_1^{(1)}(\cos\theta) = -\sin\theta \tag{5.15c}$$

$$P_1^{(-1)}(\cos\theta) = \frac{1}{2}\sin\theta \qquad (5.15d)$$

The $P_n^{(m)}(z)$ and $P_n^{(-m)}(z)$ functions are linked by the relationship

$$P_n^{(-m)}(z) = (-1)^m \frac{(n-m)!}{(n+m)!} P_n^{(m)}(z) \qquad (5.16)$$

The derivative of the associated Legendre function can be calculated by the expression

$$\frac{dP_n^{(m)}(z)}{dz} = \frac{1}{1-z^2}\left[(n+m)P_{n-1}^{(m)}(z) - nzP_n^{(m)}(z)\right] \qquad (5.17)$$

The associated Legendre functions are linked to the Legendre polynomial by the equality

$$P_n^{(m)}(z) = (-1)^m \left(1-z^2\right)^{\frac{m}{2}} \frac{d^m}{dz^m} P_n^{(0)}(z) \qquad (5.18)$$

The associated Legendre functions corresponding to various values of n and m are linked by recurrence formulae. Let us write two of them that will come in handy later:

$$zP_n^{(m)}(z) = \frac{1}{2n+1}\left[(n-m+1)P_{n+1}^{(m)}(z) + (n+m)P_{n-1}^{(m)}(z)\right] \qquad (5.19)$$

$$P_{n-1}^{(m)}(z) - P_{n+1}^{(m)}(z) = (2n+1)\sqrt{1-z^2}\, P_n^{(m-1)}(z) \qquad (5.20)$$

Consider some integral values of the associated Legendre functions

$$\int_{-1}^{1} P_n^{(m)}(x)P_l^{(m)}dx = 0, \quad l \neq n \qquad (5.21a)$$

$$\int_{0}^{\pi} P_n^{(m)}(\cos\theta)P_l^{(m)}(\cos\theta)\sin\theta d\theta = 0, \quad l \neq n \qquad (5.21b)$$

$$\int_{-1}^{1} P_n^{(m)}(x)P_n^{(l)}(x)(1-x^2)^{-1}dx = 0, \quad l \neq m \qquad (5.22a)$$

$$\int_{0}^{\pi} P_n^{(m)}(\cos\theta)P_n^{(l)}(\cos\theta)(\sin\theta)^{-1}d\theta = 0, \quad l \neq m \qquad (5.22b)$$

$$\int_{-1}^{1} \left(P_n^{(m)}(x)\right)^2 dx = \frac{2}{2n+1}\frac{(n+m)!}{(n-m)!} \qquad (5.23a)$$

$$\int\limits_{0}^{\pi}\left(P_n^{(m)}(\cos\theta)\right)^2 \sin\theta d\theta = \frac{2}{2n+1}\frac{(n+m)!}{(n-m)!} \tag{5.23b}$$

$$\int\limits_{-1}^{1}\left(P_n^{(m)}(x)\right)^2 (1-x^2)^{-1} dx = \frac{(n+m)!}{|m|(n-m)!}, \; m \neq 0 \tag{5.24a}$$

$$\int\limits_{0}^{\pi}\left(P_n^{(m)}(\cos\theta)\right)^2 (\sin\theta)^{-1} d\theta = \frac{(n+m)!}{|m|(n-m)!}, \; m \neq 0 \tag{5.24b}$$

5.3   Calculating the Radiant Power of Spherical Harmonics

The radiant electric and magnetic fields of the waves of various types are determined by expressions (5.5) and (5.7) with substitution of the $h_n^{(2)}(kr)$ Hankel function of the second kind for the spherical Bessel function $j_n(kr)$ of the first kind in them. Let us calculate the actual radiated power associated with each wave type. We engage in these computations for a number of reasons. First, this is done for a closer examination of the properties of the spherical Bessel and associated Legendre functions. Second, during the computations we obtain the integral, which will be useful later in other mathematical manipulations. Third, power normalization of the associated Legendre functions is utilized in estimating the limit capacity of a three-dimensional wireless channel during statistical modeling.

For computing the actual radiant power P it is necessary to calculate the real part of the Poynting flux vector in terms of the sphere surface of radius r [23]

$$P = \text{Re} \int\limits_{0}^{2\pi} \int\limits_{0}^{\pi} \Pi_r r^2 \sin\theta d\theta d\varphi \tag{5.25}$$

The radial component Π_r of the Poynting vector is determined in terms of fields by the formula [23]

$$\Pi_r = \frac{1}{2}(E_\theta H_\varphi^* - E_\varphi H_\theta^*) \tag{5.26}$$

The asterisk symbol $*$ in (5.26) denotes a complex conjugate.

Let us first compute the radiant power for the E-waves. We take the coefficient $\alpha_{n,m}$ equal to unity to make the formulae more concise. Substitution of the E-wave field (5.5) into (5.26) gives the expression

$$\Pi_r = \frac{i}{2(kr)^2} \sqrt{\frac{\varepsilon}{\mu}} \frac{d\left(krh_n^{(2)}(kr)\right)}{d(kr)} \left(krh_n^{(2)}(kr)\right)^* \times$$

$$\times \left[\left(\frac{dP_n^{(m)}(\cos\theta)}{d\theta} \right)^2 + \frac{m^2 \left(P_n^{(m)}(\cos\theta) \right)^2}{(\sin\theta)^2} \right]$$

Substituting Π_r into (5.25) and taking into account that integration with respect to φ resolves itself into multiplication by 2π, we obtain

$$P = \mathrm{Re} \left(\frac{i\pi}{k^2} \sqrt{\frac{\varepsilon}{\mu}} \frac{d\left(krh_n^{(2)}(kr)\right)}{d(kr)} \left(krh_n^{(2)}(kr)\right)^* \right) \times \tag{5.27}$$

$$\times \int_0^\pi \sin\theta \left[\left(\frac{dP_n^{(m)}(\cos\theta)}{d\theta} \right)^2 + \frac{m^2 \left(P_n^{(m)}(\cos\theta) \right)^2}{(\sin\theta)^2} \right] d\theta$$

Let us denote the integral in (5.27) by I for brevity and substitute $z = \cos\theta$ for the variable θ to evaluate it. Then, the integral can be written as

$$I = \int_{-1}^1 \left[\left(1 - z^2\right) \left(\frac{dP_n^{(m)}(z)}{dz^2} \right)^2 + \frac{m^2}{1 - z^2} \left(P_n^{(m)}(z) \right)^2 \right] dz \tag{5.28}$$

The integral of the addend in (5.28) in accordance with (5.24a) equals

$$\int_{-1}^1 \frac{m^2}{1 - z^2} \left(P_n^{(m)}(z) \right)^2 dz = |m| \frac{(n+m)!}{(n-m)!} \tag{5.29}$$

We use expression (5.17) to evaluate the integral of the augend, and rearrange the addend of (5.17) by the formula (5.19). The integrand then may be written as

$$\left(1 - z^2\right) \left(\frac{dP_n^{(m)}(z)}{dz} \right)^2 = \frac{1}{1 - z^2} \left(\frac{(n+1)(n+m)}{2n+1} P_{n-1}^{(m)}(z) - \frac{n(n-m+1)}{2n+1} P_{n+1}^{(m)}(z) \right)^2$$

For manipulating the result of raising the product $P_{n-1}^{(m)}(z)\, P_{n+1}^{(m)}(z)$ to the second power, let us square (5.20). We may write the obtained expression as

$$-2P_{n-1}^{(m)}(z)P_{n+1}^{(m)}(z) = (2n+1)^2 (1 - z^2)\left(P_n^{(m-1)}(z)\right)^2 - \left(P_{n+1}^{(m)}(z)\right)^2 - \left(P_{n-1}^{(m)}(z)\right)^2$$

The written above equality permits us to recast the integrand to give

$$\left(1 - z^2\right) \left(\frac{dP_n^{(m)}(z)}{dz} \right)^2 = \frac{m(n+1)(n+m)}{(1 - z^2)(2n+1)} \left(P_{n-1}^{(m)}(z) \right)^2 -$$

$$-\frac{mn(n-m+1)}{(1-z^2)(2n+1)}\left(P_{n+1}^{(m)}(z)\right)^2 + n(n+1)(n+m)(n-m+1)\left(P_n^{(m-1)}(z)\right)^2$$

Formulae (5.23a) and (5.24a) enable us to compute the value of the integral of the augend in (5.28)

$$\int_{-1}^{1}(1-z^2)\left(\frac{dP_n^{(m)}(z)}{dz}\right)^2 dz = \frac{2n^2-2|m|n+2n-|m|}{2n+1}\frac{(n+m)!}{(n-m)!}$$

Adding the resulting expression to the value of the integral of the addend (5.29), we have the formula for the integral I, included in formula (5.27) for numerical evaluation of the actual radiant power of a harmonic

$$\int_{-1}^{1}\left[(1-z^2)\left(\frac{dP_n^{(m)}(z)}{dz}\right)^2 + \frac{m^2}{1-z^2}\left(P_n^{(m)}(z)\right)^2\right]dz = \frac{2n(n+1)}{2n+1}\frac{(n+m)!}{(n-m)!} \qquad (5.30)$$

In accordance with (5.27) for actual radiant power of a harmonic we get

$$P = \mathrm{Re}\left(\frac{i\pi}{k^2}\sqrt{\frac{\varepsilon}{\mu}}\frac{d\left(krh_n^{(2)}(kr)\right)}{d(kr)}\left(krh_n^{(2)}(kr)\right)^*\right)\frac{2n(n+1)}{2n+1}\frac{(n+m)!}{(n-m)!}$$

Assuming that the sphere radius r is infinitely large we apply asymptotic representation (5.11) to the $h_n^{(2)}(kr)$ function. We then obtain the final expression for the actual radiant power of a spherical E-harmonic

$$P = \frac{\pi}{k^2}\sqrt{\frac{\varepsilon}{\mu}}\frac{2n(n+1)}{2n+1}\frac{(n+m)!}{(n-m)!} \qquad (5.31)$$

Let us calculate the actual power radiated by a spherical H-harmonic. For brevity's sake, we take the coefficient $\beta_{n,m}$ equal to 1. Substituting expressions (5.7) into (5.26) and replacing the functions $j_n(kr)$ with $h_n^{(2)}(kr)$ we obtain

$$\Pi_r = \frac{1}{2ikr}\sqrt{\frac{\mu}{\varepsilon}}h_n^{(2)}(kr)\left(\frac{d\left(krh_n^{(2)}(kr)\right)}{d(kr)}\right)^*\left[\left(\frac{dP_n^{(m)}(\cos\theta)}{d\theta}\right)^2 + \frac{m^2\left(P_n^{(m)}(\cos\theta)\right)^2}{(\sin\theta)^2}\right]$$

Insertion of Π_r in (5.25) gives for the H-wave

$$P = \mathrm{Re}\left(-\frac{i\pi r}{k}\sqrt{\frac{\mu}{\varepsilon}}h_n^{(2)}(kr)\left(\frac{d\left(krh_n^{(2)}(kr)\right)}{d(kr)}\right)^* \right) \times$$

$$\times \int_0^\pi \sin\theta\left[\left(\frac{dP_n^{(m)}(\cos\theta)}{d\theta}\right)^2 + \frac{m^2\left(P_n^{(m)}(\cos\theta)\right)^2}{(\sin\theta)^2} \right]d\theta$$

Using the integral expression (5.30) derived earlier and asymptotic expression (5.11) for the function $h_n^{(2)}(kr)$, we have the ultimate expression for the actual power radiated by a spherical H-harmonic

$$P = \frac{\pi}{k^2}\sqrt{\frac{\mu}{\varepsilon}}\frac{2n(n+1)}{2n+1}\frac{(n+m)!}{(n-m)!} \tag{5.32}$$

In what follows, expressions (5.31, 5.32) will be used in the normalization of spherical harmonics. The normalizing factor for the associated Legendre functions will be selected so that the radiant power of each spherical harmonic is unity.

5.4   Numerical Evaluation of Coefficients for Expansion of Electromagnetic Fields into Spherical Harmonics

The functions

$$P_n^{(m)}(\cos\theta)e^{im\varphi} \ , \ n = 0,1,2,\ldots; \ m = \text{-}n,\text{-}(n\text{-}1),\ldots,(n\text{-}1) \tag{5.33}$$

form an orthogonal system of functions on the sphere [18] for

$$0 \le \theta \le \pi, \ 0 \le \varphi \le 2\pi \tag{5.34}$$

These are called spherical harmonics. Therefore, if the values of the U (or V) potentials on some sphere of radius r_0 are known, the quantities $\alpha_{n,m}$ $(\beta_{n,m})$ can be worked out as the coefficients in an orthogonal basis function expansion, namely to determine $\alpha_{n,m}$ multiply (5.2) by

$$\sin\theta P_n^{(m)}(\cos\theta)e^{-im\varphi}$$

and integrate the resulting expression with respect to the domain of the spherical harmonics (5.34). In terms of the orthogonality and equality (5.23b) we obtain:

$$\alpha_{n,m} = \frac{k}{r_0 j_n(kr_0)}\frac{2n+1}{4\pi}\frac{(n-m)!}{(n+m)!}\int_0^{2\pi}\int_0^\pi U(r_0,\theta,\varphi)P_n^{(m)}(\cos\theta)e^{-im\varphi}\sin\theta d\theta d\varphi \tag{5.35}$$

A similar expression can be used for numerical evaluation of the coefficients $\beta_{n,m}$ from the potential V. Though expression (5.35) defines the coefficients $\alpha_{n,m}$ ($\beta_{n,m}$), its use in actual computations presents a real challenge. In actual practice, as a rule, electromagnetic fields are known, and not the potentials. Due to this, let us see how the coefficients $\alpha_{n,m}$ ($\beta_{n,m}$) can be determined if there is a known field in some neighborhood of the coordinate origin, or if a far zone radiation field is known (a complex pattern).

It is assumed that an electric field, including no field sources, is known in a certain neighborhood of the coordinate origin. In order to determine the coefficients $\alpha_{n,m}$ evaluate the following integral

$$A = \int_0^{2\pi} \int_0^{\pi} P_n^{(m)}(\cos\theta)e^{-im\varphi}\left[\frac{\partial(\sin\theta E_\theta)}{\partial\theta} + \frac{\partial E_\varphi}{\partial\varphi}\right]d\theta d\varphi \tag{5.36}$$

It is assumed that the fields E_θ and E_φ found in (5.36) are written as a total of the summands of the form (5.5) and (5.7), corresponding to various values of $n1$ and $m1$. The summation indices are labeled $n1$ and $m1$ to distinguish them from the fixed indices n and m in (5.36). First and foremost, notice that the expression in square brackets (5.36) vanishes for H-waves (5.7). For this reason, only the fields of E-waves are taken into account in (5.36). Substitution of (5.5) into (5.36) and integration with respect to φ in terms of the orthogonality of the $e^{im\varphi}$ functions gives:

$$A = \frac{2\pi}{kr} \int_0^{\pi} P_n^{(m)}(\cos\theta) \sum_{n1=0}^{\infty} \alpha_{n1,m} \frac{d\left(krj_{n1}(kr)\right)}{d(kr)} \times$$

$$\times \left[\frac{d\left(\sin\theta \frac{d\left(P_{n1}^{(m)}(\cos\theta)\right)}{d\theta}\right)}{d\theta} - \frac{m^2 P_{n1}^{(m)}(\cos\theta)}{\sin\theta}\right]d\theta$$

Let us change the integration and summation procedure and use the integration-by-parts formula in evaluating the integral of the augend. We obtain:

$$A = -\frac{2\pi}{kr} \sum_{n1=0}^{\infty} \alpha_{n1,m} \frac{d\left(krj_{n1}(kr)\right)}{d(kr)} \int_0^{\pi} \left[\sin\theta \frac{dP_n^{(m)}(\cos\theta)}{d\theta}\frac{dP_{n1}^{(m)}(\cos\theta)}{d\theta} + \frac{m^2 P_n^{(m)}(\cos\theta)P_{n1}^{(m)}(\cos\theta)}{\sin\theta}\right]d\theta \tag{5.37}$$

In (5.37) for $n1 = n$ the value of the integral was calculated during radiant power computations for a spherical harmonic (5.30). For $n1 \neq n$ this integral corresponds to the common radiant power of the harmonics with indices $n1,m$ and n,m. The common power is zero due to the orthogonality of the spherical harmonics. Therefore, the quantity A (5.37) in terms of (5.30) equals:

$$A = -\alpha_{n,m} \frac{4\pi}{kr} \frac{d\left(krj_n(kr)\right)}{d(kr)} \frac{n(n+1)}{2n+1} \frac{(n+m)!}{(n-m)!}$$

Hence, taking into account the designations of (5.35) for A we obtain a computational formula for the coefficients $\alpha_{n,m}$

$$\alpha_{n,m} = \frac{-kr \dfrac{2n+1}{n(n+1)} \dfrac{(n-m)!}{(n+m)!}}{4\pi \dfrac{d\left(krj_n(kr)\right)}{d(kr)}} \int_0^{2\pi}\int_0^{\pi} P_n^{(m)}(\cos\theta)e^{-im\varphi}\left[\frac{\partial\left(\sin\theta E_\theta\right)}{\partial\theta} + \frac{\partial E_\varphi}{\partial\varphi}\right]d\theta d\varphi \tag{5.38}$$

Calculating the limit of the right-hand member (38) with the radius $r \to 0$, we may derive a simpler expression for the coefficients. In terms of (5.13) and (5.12) the derivative of the spherical Bessel function included in (5.38) can be written as

$$\frac{d\left(krj_n(kr)\right)}{d(kr)} = (kr)^n \frac{n+1}{1\cdot3\cdot5\cdot\ldots\cdot(2n+1)} + 0(kr)^{n+2} \tag{5.39}$$

Here $0(kr)^{n+2}$ denotes the quantity kr, whose power is no less than $n+2$. In terms of (5.39) we have a final expression for numerical computations of the coefficients $\alpha_{n,m}$

$$\alpha_{n,m} = -(1\cdot3\cdot5\cdot\ldots\cdot(2n+1))\frac{2n+1}{4\pi n(n+1)^2} \frac{(n-m)!}{(n+m)!} \times$$

$$\times \lim_{kr\to0}\frac{1}{(kr)^{n-1}}\int_0^{2\pi}\int_0^{\pi} P_n^{(m)}(\cos\theta)e^{-im\varphi}\left[\frac{\partial\left(\sin\theta E_\theta\right)}{\partial\theta} + \frac{\partial E_\varphi}{\partial\varphi}\right]d\theta d\varphi \tag{5.40}$$

For calculation of the coefficients $\beta_{n,m}$ let us evaluate the following integral

$$B = \int_0^{2\pi}\int_0^{\pi} P_n^{(m)}(\cos\theta)e^{-im\varphi}\left[\frac{\partial E_\theta}{\partial\varphi} - \frac{\partial\left(\sin\theta E_\varphi\right)}{\partial\theta}\right]d\theta d\varphi \tag{5.41}$$

In (5.41) the bracketed expression vanishes with substitution of the E-wave fields (5.5) into it. Insertion of the H-waves fields (5.7) and manipulations similar to those performed in integral evaluation (5.36) yield

$$B = \beta_{n,m}4i\pi\sqrt{\frac{\mu}{\varepsilon}}j_n(kr)\frac{n(n+1)}{2n+1}\frac{(n+m)!}{(n-m)!} \tag{5.42}$$

Hence we get the calculation formula for the coefficients $\beta_{n,m}$

$$\beta_{n,m} = \frac{\sqrt{\dfrac{\varepsilon}{\mu} \dfrac{2n+1}{n(n+1)} \dfrac{(n-m)!}{(n+m)!}}}{4\pi i j_n (kr)} \times$$

$$\times \int\limits_0^{2\pi} \int\limits_0^\pi P_n^{(m)}(\cos\theta) e^{-im\varphi} \left[\frac{\partial \left(E_\theta\right)}{\partial\varphi} - \frac{\partial \left(\sin\theta E_\varphi\right)}{\partial\theta} \right] d\theta d\varphi \qquad (5.43)$$

Use of expression (5.12) results in a final formula for the coefficients $\beta_{n,m}$

$$\beta_{n,m} = (1\cdot 3\cdot 5\cdot\ldots\cdot(2n+1))\frac{2n+1}{4\pi i n(n+1)}\frac{(n-m)!}{(n+m)!}\sqrt{\frac{\varepsilon}{\mu}}\times$$

$$\times \lim_{kr\to 0}\frac{1}{(kr)^n}\int\limits_0^{2\pi}\int\limits_0^\pi P_n^{(m)}(\cos\theta)e^{-im\varphi}\left[\frac{\partial\left(E_\theta\right)}{\partial\varphi}-\frac{\partial\left(\sin\theta E_\varphi\right)}{\partial\theta}\right]d\theta d\varphi \qquad (5.44)$$

While working out a real problem and determining the expansion coefficients of a radiation field, formulae (5.40) and (5.44) should be modified. They have been presented in [17] and are of the form

$$\alpha_{n,m} = \frac{1}{4\pi i^{n-2}}\frac{2n+1}{n(n+1)}\frac{(n-m)!}{(n+m)!}\times$$

$$\times \lim_{kr\to\infty}\int\limits_0^{2\pi}\int\limits_0^\pi P_n^{(m)}(\cos\theta)e^{-im\varphi}\left[\frac{\partial\left(\sin\theta E_\theta e^{ikr} kr\right)}{\partial\theta}+\frac{\partial\left(E_\varphi e^{ikr} kr\right)}{\partial\varphi}\right]d\theta d\varphi \qquad (5.45)$$

$$\beta_{n,m} = \frac{1}{4\pi i^{n-2}}\frac{2n+1}{n(n+1)}\frac{(n-m)!}{(n+m)!}\sqrt{\frac{\varepsilon}{\mu}}\times$$

$$\times \lim_{kr\to\infty}\int\limits_0^{2\pi}\int\limits_0^\pi P_n^{(m)}(\cos\theta)e^{-im\varphi}\left[\frac{\partial\left(E_\theta e^{ikr} kr\right)}{\partial\varphi}-\frac{\partial\left(\sin\theta E_\varphi e^{ikr} kr\right)}{\partial\theta}\right]d\theta d\varphi \qquad (5.46)$$

The far zone expressions $E_\theta e^{ikr} kr$, $E_\varphi e^{ikr} kr$ are r-independent and represent the complex patterns of a radiating element (an antenna). Thus, formulae (5.45, 5.46) permit determination of the coefficients of spherical harmonics from the pattern.

5.5 Expansion of a Dipole Field into Spherical Harmonics

By way of illustrating the use of the formulae from the previous section, let us calculate the coefficients for expansion of an electric dipole field into spherical harmonics. The dipole is assumed to be aligned with the z axis of a spherical coordinate system. The components of the dipole electric field in a far zone appear as [23]:

$$E_\theta = -\frac{pk^2}{4\pi\varepsilon r}e^{-ikr}\sin\theta \, , \; E_\varphi = 0 \qquad (5.47)$$

Here p denotes a dipole moment. Substitution of (5.47) into (5.45) yields

$$\alpha_{n,m} = \frac{1}{4\pi i^{n-2}}\frac{2n+1}{n(n+1)}\frac{(n-m)!}{(n+m)!}\times$$

$$\times \int\limits_0^{2\pi}\int\limits_0^\pi P_n^{(m)}(\cos\theta)e^{-im\varphi}\left[-\frac{pk^3}{4\pi\varepsilon}\frac{d(\sin\theta)^2}{d\theta}\right]d\theta d\varphi$$

The integral over φ is different from zero only for $m = 0$ and equals 2π. Write

$$\alpha_{n,0} = \frac{pk^3}{4\pi i^n \varepsilon}\frac{2n+1}{n(n+1)}\int\limits_0^\pi\cos\theta\sin\theta P_n^{(0)}(\cos\theta)d\theta$$

In terms of (5.15)

$$\cos\theta = P_1^{(0)}(\cos\theta)$$

Equalities (5.21b) and (5.23b) then give:

$$\alpha_{1,0} = \frac{pk^3}{4\pi i\varepsilon} \, , \; \alpha_{n,m} = 0 \, , \text{ for } n \neq 1, \, m \neq 0 \qquad (5.48)$$

Substitution of fields (5.47) into expression (5.46) gives zero values for the coefficients $\beta_{n,m}$. The similar results for expansion of dipole field into spherical harmonics were achieved in [24].

The dipole electric and magnetic potentials are thus equal:

$$U = \frac{pk}{4\pi i\varepsilon}krh_1^{(2)}(kr)P_1^{(0)}(\cos\theta) \, , \; V = 0$$

Taking into account (5.10) for $h_1^{(2)}(kr)$ and (5.15) for $P_1^{(0)}(kr)$ produces the final expression of the dipole electric potential

$$U = \frac{pki}{4\pi\varepsilon}e^{-ikr}\left(1+\frac{1}{ikr}\right)\cos\theta \qquad (5.49)$$

Substitution (5.49) into (5.4) permits us to write all the components of the dipole electric and magnetic fields as:

$$E_r = \frac{pki}{2\pi \varepsilon r^2} e^{-ikr} \left(1 + \frac{1}{ikr}\right) \cos\theta$$

$$E_\theta = \frac{p}{4\pi \varepsilon r} e^{-ikr} \left(-k^2 + \frac{ik}{r} + \frac{1}{r^2}\right) \sin\theta$$

$$E_\varphi = 0$$

$$H_r = 0$$

$$H_\theta = 0$$

$$H_\varphi = \frac{p}{4\pi \varepsilon r} \sqrt{\frac{\varepsilon}{\mu}} e^{-ikr} \left(k^2 - \frac{ik}{r}\right) \sin\theta$$

The resulting expressions coincide with the dipole field equations of [23]. Furthermore, in addition to the far zone fields, the coincidence extends to the closest area too. Such coincidence is not accidental. Employment of the $h_n^{(2)}(kr)$ function in lieu of its asymptotic representation (5.11) defines an analytical extension of the potential to the domain of small kr values and permits determination of the near zone field of a radiator from its far zone field [17].

5.6 Expansion of a Plane Electromagnetic Wave into Spherical Harmonics (downfall)

It is common practice in channel modeling to represent a field in the receiving area as a sum of diversely directed plane waves. Consider a problem of a plane wave expansion into spherical harmonics. The solution of this problem could be found in [24,25]. In those books the expansion of the plane electromagnetic wave into spherical harmonics was used to solve the problem of diffraction of the electromagnetic waves on the sphere. A solution to this problem provides a basis for estimating channel characteristics. In this section we concern ourselves with expansion of a wave propagating along the z axis. An arbitrarily directed wave is being considered in Sect. 8.

The electric and magnetic fields of a plane wave propagating from an infinitely distant point of axis z are shown schematically in Fig. 5.4.

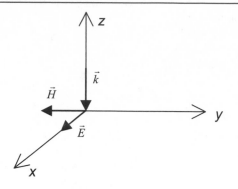

Fig. 5.4. Downfall of a plane electromagnetic wave

The Cartesian components of the electric field are:

$$E_x = E_0 e^{ikr\cos\theta}, \quad E_y = 0, \quad E_z = 0 \tag{5.50}$$

In order to calculate the projections E_θ and E_φ let us write the expressions for the unit vectors ($\mathbf{r}_0$, θ_0, φ_0) of the spherical coordinate system

$$\mathbf{r}_0 = \mathbf{x}_0 \sin\theta\cos\varphi + \mathbf{y}_0 \sin\theta\sin\varphi + \mathbf{z}_0 \cos\theta \tag{5.51a}$$

$$\theta_0 = \mathbf{x}_0 \cos\theta\cos\varphi + \mathbf{y}_0 \cos\theta\sin\varphi - \mathbf{z}_0 \sin\theta \tag{5.51b}$$

$$\varphi_0 = -\mathbf{x}_0 \sin\varphi + \mathbf{y}_0 \cos\varphi \tag{5.51c}$$

Computations of the **E**-field scalar product (5.50) and unit vectors (5.51) allow us to write the E_θ and E_φ components of the electric field

$$E_\theta = \cos\theta\cos\varphi e^{ikr\cos\theta} \tag{5.52a}$$

$$E_\varphi = -\sin\varphi e^{ikr\cos\theta} \tag{5.52b}$$

In (5.52) and the subsequent formulae E_0 is taken equal to unity for brevity of writing.

Let us use formulae (5.40) and (5.44) for computations of the coefficients of field expansion (5.52) into spherical harmonics. With substitution of (5.52) the bracketed expression in formula (5.40) becomes:

$$\frac{\partial\left(\sin\theta E_\theta\right)}{\partial\theta} + \frac{\partial E_\varphi}{\partial\varphi} = -\left(2 + ikr\cos\theta\right)\sin^2\theta\cos\varphi e^{ikr\cos\theta} \tag{5.53}$$

Let us evaluate the double integral included in (5.40) in terms of expression (5.53). Integration over φ of the φ–dependent functions gives:

$$\int_0^{2\pi} \cos\varphi\, e^{-im\varphi}\, d\varphi = \pi, \quad \text{for } |m| = 1$$

and 0 for all the rest values of m. To calculate the value of the integral in (5.40) we write the exponential function series

$$e^{ikr\cos\theta} = \sum_{l=0}^{\infty} \frac{1}{l!}\left(ikr\cos\theta\right)^l \tag{5.54}$$

and replace the integration variable θ with $z = \cos\theta$. For $m = 1$ we have

$$\int_0^{2\pi}\int_0^{\pi} P_n^{(1)}(\cos\theta)e^{-i\varphi}\left[\frac{\partial\left(\sin\theta E_\theta\right)}{\partial\theta} + \frac{\partial E_\varphi}{\partial\varphi}\right]d\theta d\varphi =$$

$$= -\pi \int_{-1}^{1} \sqrt{1-z^2}\, P_n^{(1)}(z) \sum_{l=0}^{\infty} \frac{\left(ikrz\right)^l\left(2+ikrz\right)}{l!}\, dz$$

Further we use equality (5.18), which for $m = 1$ assumes the form:

$$P_n^{(1)}(z) = -\sqrt{1-z^2}\,\frac{d}{dz}P_n^{(0)}(z) \tag{5.55}$$

Employment of (5.55) and integration by parts results in the following expression of the integral for $m = 1$

$$-\pi\int_{-1}^{1} P_n^{(0)}(z)\frac{d}{dz}\left(\left(1-z^2\right)\sum_{l=0}^{\infty}\frac{\left(ikrz\right)^l\left(2+ikrz\right)}{l!}\right)dz \tag{5.56}$$

From expression (5.40) it will be obvious that while evaluating the integrals of (5.56), the summands containing kr raised to the power in excess of $n-1$, can be omitted. These summands add nothing to the limit value in (5.40). Consideration will be also given to the fact, known from the Legendre polynomial theory [19], that

$$\int_{-1}^{1} P_n^{(0)}(z)f(z)dz = 0 \tag{5.57}$$

if $f(z)$ is a polynomial whose power is under n. Then (5.56) may be written as

$$\pi(ikr)^{n-1}\frac{(n+1)^2}{(n-1)!}\int_{-1}^{1} z^n P_n^{(0)}(z)dz \tag{5.58}$$

For evaluation of the integral in (5.58) let us use (5.19), which with a zero upper index of the Legendre function ($m = 0$) assumes the form

$$zP_n^{(0)}(z) = \frac{n}{2n+1}P_{n-1}^{(0)}(z) + \frac{n+1}{2n+1}P_{n+1}^{(0)}(z)$$

The addend in this formula can be omitted, as it contributes nothing to the value of the integral (5.58) due to property (5.57). Use of the written expression n times permits us to evaluate the integral in (5.58)

$$\int_{-1}^{1} z^n P_n^{(0)}(z)dz = 2\frac{n!}{1\cdot3\cdot5\cdot\ldots\cdot(2n+1)} \tag{5.59}$$

Thus, the double integral in (5.40), determined by expression (5.58), for $m = 1$ equals

$$\int_0^{2\pi}\int_0^{\pi} P_n^{(1)}(\cos\theta)e^{-i\varphi}\left[\frac{\partial(\sin\theta E_\theta)}{\partial\theta} + \frac{\partial E_\varphi}{\partial\varphi}\right]d\theta d\varphi \cong 2\pi(ikr)^{n-1}\frac{n(n+1)^2}{1\cdot3\cdot5\cdot\ldots\cdot(2n+1)} \tag{5.60a}$$

With $m = -1$ similar transformations result in a value differing from (5.60a) by the multiplication factor $(-1)n^{-1}(n+1)^{-1}$. Such conclusion follows from comparing (5.54) for $m = 1$ and for $m = -1$. Presence of this multiplication factor also follows from the relationship of $P_n^{(-1)}$ and $P_n^{(1)}$, defined by the equality (5.16).

$$\int_0^{2\pi}\int_0^{\pi} P_n^{(-1)}(\cos\theta)e^{i\varphi}\left[\frac{\partial(\sin\theta E_\theta)}{\partial\theta} + \frac{\partial E_\varphi}{\partial\varphi}\right]d\theta d\varphi \cong$$

$$\cong -2\pi(ikr)^{n-1}\frac{n+1}{1\cdot3\cdot5\cdot\ldots\cdot(2n+1)} \tag{5.60b}$$

Substitution of (5.60) into (5.40) determines the coefficients $\alpha_{n,m}$ for the plane wave:

$$\alpha_{n,1} = -i^{n-1}\frac{2n+1}{2n(n+1)} \qquad \alpha_{n,-1} = i^{n-1}\frac{2n+1}{2} \tag{5.61}$$

$$\alpha_{n,m} = 0 \text{ , for } |m| \neq 1$$

Let us turn now to numerical computation of the coefficients $\beta_{n,m}$, defining the H-wave fields. For this we call the reader's attention to formula (5.44). The bracketed expression (5.44) with substitution of (5.52) into it rearranges to the form of

$$\frac{\partial E_\theta}{\partial\varphi} - \frac{\partial(\sin\theta E_\varphi)}{\partial\theta} = -ikr\sin^2\theta\sin\varphi e^{ikr\cos\theta} \tag{5.62}$$

Substitute (5.62) in (5.44) and integrate with respect to φ

$$\int_0^{2\pi} \sin\varphi e^{-im\varphi} d\varphi = \begin{cases} -i\pi......m = 1 \\ i\pi.........m = -1 \\ 0...........|m| \neq 1 \end{cases}$$

Use of the exponential function series and changing to the variable $z = \cos\theta$ gives the following expression of the integral in (5.44) for $m = 1$

$$\int_0^{2\pi}\int_0^{\pi} P_n^{(1)}(\cos\theta)e^{-i\varphi}\left[\frac{\partial(\sin\theta E_\theta)}{\partial\theta} - \frac{\partial E_\varphi}{\partial\varphi}\right]d\theta d\varphi =$$

$$= -\pi \int_{-1}^{1}\sqrt{1-z^2}\, P_n^{(1)}(z) \sum_{l=0}^{\infty} \frac{(ikr)^{l+1} z^l}{l!} dz$$

Equality (5.55) and integration by parts permits writing the integral as

$$-\pi kr \int_{-1}^{1} P_n^{(0)}(z)\frac{d}{dz}\left((1-z^2)\sum_{l=0}^{\infty}\frac{(ikrz)^l}{l!}\right)dz \qquad (5.63)$$

Expression (5.44) is the evidence that in (5.63) we may disregard the summands that contain kr of power greater than n. Taking into account (5.57) and (5.59) permits evaluation of the integral of (5.63)

$$\int_0^{2\pi}\int_0^{\pi} P_n^{(1)}(\cos\theta)e^{-i\varphi}\left[\frac{\partial E_\theta}{\partial\varphi} - \frac{\partial(\sin\theta E_\varphi)}{\partial\theta}\right]d\theta d\varphi \cong 2\pi kr(ikr)^{n-1}\frac{n(n+1)}{1\cdot3\cdot5\cdot...\cdot(2n+1)}$$

With $m = -1$ similar calculations yield

$$\int_0^{2\pi}\int_0^{\pi} P_n^{(-1)}(\cos\theta)e^{i\varphi}\left[\frac{\partial E_\theta}{\partial\varphi} - \frac{\partial(\sin\theta E_\varphi)}{\partial\theta}\right]d\theta d\varphi \cong 2\pi kr(ikr)^{n-1}\frac{1}{1\cdot3\cdot5\cdot...\cdot(2n+1)}$$

Substitution of the integral values into (5.44) defines the coefficients $\beta_{n,m}$:

$$\beta_{n,1} = -i^n\sqrt{\frac{\varepsilon}{\mu}\frac{2n+1}{2n(n+1)}} \quad \beta_{n,-1} = -i^n\sqrt{\frac{\varepsilon}{\mu}\frac{2n+1}{2}} \qquad (5.64)$$

$$\beta_{n,m} = 0 \text{ , for } |m| \neq 1$$

Expressions (5.61) and (5.64) determine the coefficients of spherical harmonics for downward vertical movement of a plane wave. The arbitrary plane wave is considered in Sect. 5.8 after discussing spherical harmonics transformations during rotation of the coordinate system.

5.7   Spherical Harmonics Transformations for Rotation of the Coordinate System

Analysis of the informational properties of multipath wireless channels requires numerical evaluation of the expansion coefficients for arbitrarily directed plane waves. Use of the calculation formulae presented in Sect. 5.6 for a plane wave propagating along the z-axis involves frighteningly cumbersome expressions. This is why the transformation formulae for spherical harmonics are given in this section. Later, in Sect. 8, they will come useful during discussion of the arbitrarily directed wave.

Figure 5.5 depicts two coordinate systems, one of which differs from the other by being turned about the y axis through some angle θ_1. Rotation about the z axis through the angle φ_1 will be considered later. It is assumed that the coordinates of spherical harmonics in the old xyz coordinate system are known. It is necessary to determine the harmonic coefficients in the new coordinate system whose axes are denoted in Fig 5.5 by stroked characters.

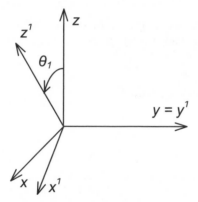

Fig. 5.5. Rotation of the coordinate system about axis y

First and foremost, let us consider how the coordinates of some point in the old coordinate system are related to those in the new system. Let us denote the new coordinates θ_2, ϕ_2, leaving the old ones θ, ϕ unchanged. The projections of a unit vector with the coordinates θ_2, ϕ_2 onto new axes are:

$$e_x^1 = \sin\theta_2 \cos\varphi_2$$

$$e_y^1 = \sin\theta_2 \sin\varphi_2$$

$$e_z^1 = \cos\theta_2$$

Taking into account the turn of the axes x^1 and z^1 through the angle θ_1, let us write the projections of the unit vector onto the old axes:

$$e_x = e_x^1 \cos\theta_1 + e_z^1 \sin\theta_1 = \sin\theta_2 \cos\varphi_2 \cos\theta_1 + \cos\theta_2 \sin\theta_1$$

$$e_y = e_y^1 = \sin\theta_2 \sin\varphi_2$$

$$e_z = -e_x^1 \sin\theta_1 + e_z^1 \cos\theta_1 = \cos\theta_2 \cos\theta_1 - \sin\theta_2 \cos\varphi_2 \sin\theta_1$$

These equalities permit expression of the old coordinates θ, ϕ in terms of the new θ_2, ϕ_2.

$$e_z = \cos\theta = \cos\theta_2 \cos\theta_1 - \sin\theta_2 \cos\varphi_2 \sin\theta_1$$

$$\frac{e_y}{e_x} = \operatorname{tg}\varphi = \frac{\sin\theta_2 \sin\varphi_2}{\sin\theta_2 \cos\varphi_2 \cos\theta_1 + \cos\theta_2 \sin\theta_1} \tag{5.65}$$

It has been demonstrated in [22] that with fulfillment of equalities (5.65) the addition theorem holds good for spherical harmonics

$$P_n^{(m)}(\cos\theta)e^{-im\varphi} = i^m \sqrt{\frac{(n+m)!}{(n-m)!}} \sum_{l=-n}^{n} i^{-l} \sqrt{\frac{(n-l)!}{(n+l)!}} e^{-il\varphi_2} P_{m,l}^{(n)}(\cos\theta_1) P_n^{(l)}(\cos\theta_2) \tag{5.66}$$

In (5.66) the function $P_{m,l}^{(n)}(z)$, dependent on three indices, is calculated by the formula

$$P_{m,l}^{(n)}(z) = i^{m-l} \sqrt{(n-m)!(n+m)!(n-l)!(n+l)!} \left(\frac{1-z}{1+z}\right)^{\frac{m-l}{2}} \left(\frac{1+z}{2}\right)^n \times$$

$$\times \sum_{j=J_1}^{J_2} \frac{(-1)^j}{j!(n-m-j)!(n+l-j)!(m-l+j)!} \left(\frac{1-z}{1+z}\right)^j \tag{5.67}$$

In (5.67) the summation limits are determined by the equalities:

$$J_1 = \max(0, l-m) \qquad J_2 = \min(n+l, n-m) \tag{5.68}$$

Let us rearrange (5.66), changing the signs of m and the summation index l. In terms of (5.16) and the equality presented in [22]

$$P_{-m,-l}^{(n)}(z) = P_{m,l}^{(n)}(z)$$

we derive

$$P_n^{(m)}(\cos\theta)e^{im\varphi} = i^m\sqrt{\frac{(n+m)!}{(n-m)!}}\sum_{l=-n}^{n}i^{-l}\sqrt{\frac{(n-l)!}{(n+l)!}}P_{m,l}^{(n)}(\cos\theta_1)P_n^{(l)}(\cos\theta_2)e^{il\varphi_2} \quad (5.69)$$

Equality (5.69) determines transformation of a spherical harmonic due to rotation of the coordinate system. However, it is obtained with the condition that the direction of the wave propagation is aligned with the xoz plane. Rewrite formula (5.69) so that it is true for the arbitrary azimuth incidence angle φ_1. To attain this, it is enough to perform the following substitution in (5.69):

$$e^{il\varphi_2} \rightarrow e^{il(\varphi_2-\varphi_1+\pi)} \quad (5.70a)$$

$$e^{im\varphi} \rightarrow e^{im(\varphi+\pi)} \quad (5.70b)$$

To make the direction to the source a benchmark for tracking the angle φ_2 the summand π has been introduced into the exponential index. With substitutions (5.70) in (5.69) we obtain

$$P_n^{(m)}(\cos\theta)e^{im\varphi} = i^{-m}\sqrt{\frac{(n+m)!}{(n-m)!}} \times$$
$$\times \sum_{l=-n}^{n}i^l\sqrt{\frac{(n-l)!}{(n+l)!}}P_{m,l}^{(n)}(\cos\theta_1)e^{-il\varphi_1}P_n^{(l)}(\cos\theta_2)e^{il\varphi_2} \quad (5.71)$$

In (5.71) θ_1 denotes the angle between the z axis and the direction to the source of a plane wave. The angle φ_1 stands for the source azimuth angle. The first member of expression (5.71) presents, as is obvious from its comparison to (5.2), the angular-coordinate dependence of the U potential. It is seen from (5.71) that one spherical E-harmonic of amplitude $\alpha_{n,m}$ with rotation of the coordinate system gives birth to $2n+1$ spherical harmonics in the new coordinate system. Their complex amplitudes $\alpha_{n,l}$ ($l = -n, -(n-1),\ldots,(n-1),n$) are therefore

$$\alpha_{n,l} = i^{l-m}\sqrt{\frac{(n+m)!\,(n-l)!}{(n-m)!\,(n+l)!}}P_{m,l}^{(n)}(\cos\theta_1)e^{-il\varphi_1}\alpha_{n,m} \quad (5.72)$$

For transformation of the $\beta_{n,m}$ coefficients of the H spherical harmonics one should use a similar expression.

5.8   Expansion of the Field of an Arbitrarily Directed Plane Wave into Spherical Harmonics

Let us use the deductions of Sect. 5.6 and 5.7 for determining the expansion coefficients of an arbitrarily directed plane wave.

In Sect. 5.6 we performed numerical evaluation of the complex amplitudes of spherical harmonics $\alpha_{n,m}$ ($\beta_{n,m}$) of a wave propagating along axis z ($\theta_1 = 0$). The wave electric field is aligned along the x axis. It is conceivable that an arbitrarily directed wave (characterized by the angles φ_1, θ_1) is derived from this wave due to turning the polarization plane about the axis z through the angle φ_1 and subsequent deflection of the propagation direction from the axis z through the angle θ_1. Then formula (5.72) can be used for calculation of the expansion coefficients. However, in the considered transformation we have a wave whose electric field lies in the plane passing through the axis z. This is a vertically polarized wave and its coefficients will be labeled by the upper index v. Later we will consider a horizontally polarized wave, the coefficients of which will be labeled by the upper index h.

The coefficients $\alpha^v_{n,l}$ are determined by the results of rearranging (5.61) for $\alpha^v_{n,m}$ in terms of (5.71). Adding the results of the rearrangements for $m = 1$ and $m = -1$ we obtain

$$\alpha^v_{n,l} = i^{n+l} \frac{2n+1}{2\sqrt{n(n+1)}} \sqrt{\frac{(n-l)!}{(n+l)!}} e^{-il\varphi_1} \left(P^{(n)}_{1,l}(\cos\theta_1) + P^{(n)}_{-1,l}(\cos\theta_1) \right) \qquad (5.73)$$

In a similar way, for the coefficients of H-waves from (5.64) and (5.72) we derive

$$\beta^v_{n,l} = i^{n+l+1} \sqrt{\frac{\varepsilon}{\mu}} \frac{2n+1}{2\sqrt{n(n+1)}} \sqrt{\frac{(n-l)!}{(n+l)!}} e^{-il\varphi_1} \left(P^{(n)}_{1,l}(\cos\theta_1) - P^{(n)}_{-1,l}(\cos\theta_1) \right) \qquad (5.74)$$

To derive the formulae for horizontal polarization, let us first consider a special transformation case ($\theta_1 = 0$, $\varphi_1 = \pi/2$). This case represents a rotation of the wave polarization plane without variations in the direction of its propagation. The direction of propagation is defined by the axis z, as shown in Fig. 5.4. Only the horizontal components of the electric field are different from 0 in the wave that results from rotation about the axis z through the angle φ_1 and deviation through angle θ_1 from the z axis. Let us call the spherical harmonic coefficients of such a wave "horizontal polarization coefficients" and label them with the upper index h. In terms of the equality presented in [22].

$$P^{(n)}_{m,l}(1) = \delta_{m,l}$$

Here $\delta_{m,l} = 1$ with $m = l$ and $\delta_{m,l} = 0$ with $m \neq l$. Then rearrangement of (5.72) for $\theta_1 = 0$, $\varphi_1 = \pi/2$ gives

$$\alpha_{n,m}^{h} = i^{-m} \alpha_{n,m}^{v} \qquad (5.75)$$

The coefficients $\beta_{n,m}$ with rotation of the polarization plane are derived in a similar manner. The expressions (5.61) and (5.64) in terms of (5.75) for vertical downward motion of the wave, whose electric field is directed along the axis y permit us to write:

$$\alpha_{n,1}^{h} = i^{n} \frac{2n+1}{2n(n+1)} \qquad \alpha_{n,-1}^{h} = i^{n} \frac{2n+1}{2} \qquad (5.76a)$$

$$\alpha_{n,m}^{h} = 0 \ \text{ for } |m| \neq 1$$

$$\beta_{n,1}^{h} = i^{n+1} \sqrt{\frac{\varepsilon}{\mu}} \frac{2n+1}{2n(n+1)} \qquad \beta_{n,-1}^{h} = -i^{n+1} \sqrt{\frac{\varepsilon}{\mu}} \frac{2n+1}{2} \qquad (5.76b)$$

$$\beta_{n,m}^{h} = 0 \ \text{ for } |m| \neq 1$$

Rotation of the coordinate system through φ_1 and θ_1 after a preliminary turn of the polarization plane through $\pi/2$, yields a horizontally polarized wave of a specified direction. Application of (5.72) to (5.76) allows derivation of an expression for the spherical harmonics coefficients of a horizontally polarized wave:

$$\alpha_{n,l}^{h} = i^{n+l-1} \frac{2n+1}{2\sqrt{n(n+1)}} \sqrt{\frac{(n-l)!}{(n+l)!}} e^{-il\varphi_1} \left(P_{1,l}^{(n)}(\cos\theta_1) - P_{-1,l}^{(n)}(\cos\theta_1) \right) \qquad (5.77)$$

$$\beta_{n,l}^{h} = i^{n+l} \sqrt{\frac{\varepsilon}{\mu}} \frac{2n+1}{2\sqrt{n(n+1)}} \sqrt{\frac{(n-l)!}{(n+l)!}} e^{-il\varphi_1} \left(P_{1,l}^{(n)}(\cos\theta_1) + P_{-1,l}^{(n)}(\cos\theta_1) \right) \qquad (5.78)$$

An arbitrarily polarized wave can be represented by a sum of the vertically and horizontally polarized waves. Thus, expressions (5.73, 5.74, 5.77 and 5.78) permit numerical evaluation of the complex coefficients of spherical harmonics for an arbitrarily directed electromagnetic plane wave.

Let us rewrite these expressions in a simpler form. For the rearrangement we use the relationships presented in [22]:

$$\sqrt{1-z^2} \frac{dP_{m,l}^{(n)}(z)}{dz} + \frac{mz-l}{\sqrt{1-z^2}} P_{m,l}^{(n)}(z) = -i\sqrt{(n-m)(n+m+1)} P_{m+1,l}^{(n)}(z) \qquad (5.79a)$$

$$\sqrt{1-z^2} \frac{dP_{m,l}^{(n)}(z)}{dz} - \frac{mz-l}{\sqrt{1-z^2}} P_{m,l}^{(n)}(z) = -i\sqrt{(n+m)(n-m+1)} P_{m-1,l}^{(n)}(z) \qquad (5.79b)$$

Assuming $m = 0$ and subtracting (5.79a) from (5.79b), we obtain

$$\frac{2l}{\sqrt{1-z^2}} P_{0,l}^{(n)}(z) = -i\sqrt{n(n+1)}\left(P_{-1,l}^{(n)}(z) - P_{1,l}^{(n)}(z)\right) \tag{5.80}$$

[22] demonstrates the relationship between the functions $P_{0,l}^{(n)}(z)$ and the associated Legendre functions $P_n^{(l)}(z)$

$$P_n^{(l)}(z) = i^l \sqrt{\frac{(n+l)!}{(n-l)!}} P_{0,l}^{(n)}(z) \tag{5.81}$$

From (5.80) and (5.81) we derive

$$P_{-1,l}^{(n)}(z) - P_{1,l}^{(n)}(z) = \frac{2li^{1-l}}{\sqrt{n(n+1)}} \sqrt{\frac{(n-l)!}{(n+l)!}} \frac{P_n^{(l)}(z)}{\sqrt{1-z^2}}$$

In terms of $z = \cos\theta$ we have the final expression for the remainder of the functions dependent on three indices and found in (5.74) and (5.77).

$$P_{-1,l}^{(n)}(\cos\theta) - P_{1,l}^{(n)}(\cos\theta) = \frac{2li^{1-l}}{\sqrt{n(n+1)}} \sqrt{\frac{(n-l)!}{(n+l)!}} \frac{P_n^{(l)}(\cos\theta)}{\sin\theta} \tag{5.82}$$

For rearrangement of the sum, let us revert to (5.79a) and (5.79b) and add them together, assuming $m = 0$.

$$2\sqrt{1-z^2} \frac{dP_{0,l}^{(n)}(z)}{dz} = -i\sqrt{n(n+1)}\left(P_{-1,l}^{(n)}(z) + P_{1,l}^{(n)}(z)\right) \tag{5.83}$$

In terms of (5.81) and $z = \cos\theta$, we have

$$P_{-1,l}^{(n)}(\cos\theta) + P_{1,l}^{(n)}(\cos\theta) = \frac{2i^{-l-1}}{\sqrt{n(n+1)}} \sqrt{\frac{(n-l)!}{(n+l)!}} \frac{d\left(P_n^{(l)}(\cos\theta)\right)}{d\theta} \tag{5.84}$$

We use (5.82) and (5.84) to bring formulae (5.73, 5.74, 5.77, 5.78) into the following form:

$$\alpha_{n,l}^v = i^{n-1} \frac{2n+1}{n(n+1)} \frac{(n-l)!}{(n+l)!} \frac{d\left(P_n^{(l)}(\cos\theta)\right)}{d\theta}\bigg|_{\theta=\theta_1} e^{-il\varphi_1} \tag{5.85}$$

$$\beta_{n,l}^v = i^n l \sqrt{\frac{\varepsilon}{\mu}} \frac{2n+1}{n(n+1)} \frac{(n-l)!}{(n+l)!} \frac{P_n^{(l)}(\cos\theta_1)}{\sin\theta_1} e^{-il\varphi_1} \tag{5.86}$$

$$\alpha_{n,l}^h = -i^n l \frac{2n+1}{n(n+1)} \frac{(n-l)!}{(n+l)!} \frac{P_n^{(l)}(\cos\theta_1)}{\sin\theta_1} e^{-il\varphi_1} \tag{5.87}$$

$$\beta_{n,l}^{h} = i^{n-1} \sqrt{\frac{\varepsilon}{\mu} \frac{2n+1}{n(n+1)} \frac{(n-l)!}{(n+l)!}} \frac{d\left(P_n^{(l)}(\cos\theta)\right)}{d\theta}\Bigg|_{\theta=\theta_1} e^{-il\varphi_1} \tag{5.88}$$

Expressions (5.85-5.88) allow numerical evaluation of the complex coefficients of spherical harmonics for an arbitrarily directed plane electromagnetic wave. These formulae are the main result of this chapter and are used in what follows in estimating the informational properties of three-dimensional wireless channels.

5.9 Summary

Chapter 5 presents some of the information about describing electromagnetic fields in terms of spherical harmonics.
1. Expressions for the electric and magnetic fields of the E- and H-harmonics have been presented.
2. Expressions for numerical evaluation of the complex amplitudes of spherical harmonics from the known electric field have been provided.
3. Formulae for numerical evaluation of the spherical harmonic complex amplitudes for the plane electromagnetic wave have been derived.

6 Matrix Element Calculation Formulae for Three-Dimensional Wireless Channels

An estimation of the limiting characteristics of a 3-D wireless channel is similar to a numerical evaluation of the characteristics of a 2-D channel. It is based on the formulae of Chap. 3. The difference resides only in the method of computing the matrix array characterizing the channel. In this chapter we present a number of calculation techniques for the three-dimensional wireless channel.

It is assumed in Sect. 6.1 that the receiver extracts information from the electromagnetic field analyzing the complex amplitudes of spherical harmonics. The matrix representation in Sect. 6.2 is based on the assumption that the receiver records the values of the electric and magnetic potentials at discrete points on the surface of a receiving area. In the matrix representation of Sect. 6.3 we operate on the premise that projections of the electric and magnetic fields are recorded at discrete points on the surface of the receiving area.

The results of a numerical evaluation of characteristics are given in Chap. 7.

6.1 Calculation of the Channel Matrix Elements in Terms of the Source Wave to Spherical Harmonics Translation

It is assumed in this section that the receiver analyzes the electromagnetic field on some spherical surface. The electromagnetic field is created by plane wave sources. The number of sources, their coordinates, polarization and the complex amplitudes of the waves produced by them may vary. It has been demonstrated in Chap. 5 that the electromagnetic field on a spherical surface can be represented by a discrete set of complex numbers, i.e. the complex amplitudes of spherical harmonics.

Thus, the channel input is thought of as a set of the complex amplitudes of plane waves. The channel output is treated as a set of the complex amplitudes of spherical harmonics. That is, the wireless channel is visualized as a linear multipole converting plane waves into spherical harmonics. The objective of this section is to derive a matrix representing this conversion, i.e. the channel matrix.

The formulae (5.85-5.88) for conversion of the complex amplitudes of plane waves to the complex amplitudes of spherical harmonics have been presented in Sect. 5.8. However, some calculation technique problems need to be solved before we are able to use the numerical evaluation relationships given there. These include the issues of spherical harmonics normalization, orderly numeration of the harmonics and wave sources, and some other issues. Let us turn to considering these matters.

First and foremost, notice that in numerical computations one should use normalized spherical harmonics. For transition from non-normalized spherical harmonics to normalized ones, introduce the multiplication factor into the calculation formulae

$$\sqrt{\frac{1}{2\pi}\frac{2n+1}{n(n+1)}\frac{(n-m)!}{(n+m)!}} \tag{6.1}$$

The need for the factor substitution (6.1) follows from the power expressions (5.31 and 5.32) for spherical harmonics. The necessity of taking into account the normalizing factor is evident from comparison of the spherical harmonic potentials with the indices n,m and $n,-m$. The field structure of these harmonics is identical and they differ by the sense of rotation through the φ angle only. Though, as is obvious from (5.16) in absence of normalization, the $P_n^{(m)}$ and $P_n^{(-m)}$ functions have materially different factors. Introduction of the normalizing factor (6.1) eliminates this non-symmetry.

Consider the normalizing factor (6.1) in the Legendre associated functions. The normalized associated Legendre functions will be denoted by $PN_n^{(m)}$

$$PN_n^{(m)}(z) = \sqrt{\frac{1}{2\pi}\frac{2n+1}{n(n+1)}\frac{(n-m)!}{(n+m)!}}P_n^{(m)}(z) \tag{6.1a}$$

Take a note that the normalizing factor (6.1) is distinct from the one more commonly used in literature [18]

$$\sqrt{\frac{2n+1}{4\pi}\frac{(n-m)!}{(n+m)!}}$$

This factor is introduced with the goal of getting the unity norm spherical harmonics, that is, from the condition

$$\int_0^{2\pi}\int_0^\pi \left(P_n^{(m)}(\cos\theta)\right)^2 \sin\theta d\theta d\varphi = 1$$

As becomes obvious from computations, the power normalization (6.1) is better for attaining the goals set out here. With the normalization (6.1) the calculation expressions assume the simplest form, and the results of various computational techniques turn out more closely aligned.

Another limitation of the calculation formulae derived in Sect. 5.8 is that the amplitudes α and β in them are of different dimensionality. This limitation can be overcome by multiplication of expressions (5.86) and (5.88) by $(\mu/\varepsilon)^{1/2}$. The $(\mu/\varepsilon)^{1/2}$ factor, i.e., the characteristic wave impedance of the medium, links the amplitudes of the electric and magnetic fields of a plane wave. The receiver is able

to extract information by picking up the wave electric or magnetic field. Removal
of the $(\varepsilon/\mu)^{1/2}$ multiplier from (5.86) and (5.88) means that detection of a plane
wave by its magnetic field is identical to its detection by the electric field.

In addition, we introduce the $j_n(kr)$ multiplier into formulae (5.85-5.88). The
spherical Bessel function $j_n(kr)$ is found as a factor in the expressions for the electric and magnetic potentials (5.2, 5.3). Let us rewrite the calculation formulae
(5.85-5.88) of the section, substituting the factor into (5.85) and (5.87)

$$j_n(kr) \tag{6.2}$$

and into (5.86) and (5.88)

$$\sqrt{\frac{\mu}{\varepsilon}} j_n(kr) \tag{6.3}$$

In addition we substitute the index designations $n1$ and $n2$ ($\alpha_{n1,n2}$ and $\beta_{n1,n2}$) for
n and m, and α and β respectively. In what follows, the values of n will be used in
continuous numbering of the spherical harmonics. The θ_l and φ_l angles determining the direction to the source of a plane wave will be denoted θ_t and φ_t. From
this point on, we will use numerical index m (source number) instead of index t in
θ_t and φ_t.

In terms of factor (6.1) and either of the factors (6.2) or (6.3), the expressions
(5.85-5.88) for the complex amplitudes of spherical harmonics with the unity amplitude of the plane wave will take on the form:

$$a_{n1,n2}^{v} = i^{n1-1} j_{n1}(kr) e^{-in2\varphi_t} \frac{d\left(PN_{n1}^{(n2)}(\cos\theta)\right)}{d\theta}\bigg|_{\theta=\theta_t} \tag{6.4a}$$

$$b_{n1,n2}^{v} = i^{n1} n2 j_{n1}(kr) e^{-in2\varphi_t} \frac{PN_{n1}^{(n2)}(\cos\theta_t)}{\sin\theta_t} \tag{6.4b}$$

$$a_{n1,n2}^{h} = -i^{n1} n2 j_{n1}(kr) e^{-in2\varphi_t} \frac{PN_{n1}^{(n2)}(\cos\theta_t)}{\sin\theta_t} \tag{6.4c}$$

$$b_{n1,n2}^{h} = i^{n1-1} j_{n1}(kr) e^{-in2\varphi_t} \frac{d\left(PN_{n1}^{(n2)}(\cos\theta)\right)}{d\theta}\bigg|_{\theta=\theta_t} \tag{6.4d}$$

In (6.4) we have used symbols $a_{n1,n2}$ and $b_{n1,n2}$ in lieu of $\alpha_{n1,n2}$ and $\beta_{n1,n2}$. The
quantities $a_{n1,n2}$ and $b_{n1,n2}$ differ form the coefficients of spherical harmonics $\alpha_{n1,n2}$
and $\beta_{n1,n2}$ by the multiplication factor $j_{n1}(kr)$.

Consider next the issue of numbering the spherical harmonics and the sources
of plane waves. The double-index notation of the spherical harmonics employed in

analytical manipulations should be replaced by a single-index designation, which is preferable in numerical evaluation.

First and foremost, notice that the spherical harmonic with the indices $n1 = 0$, $n2 = 0$ should not be taken into account during computations. The potential determined by it is not dependent on the coordinates since $P_0^{(0)}(\cos\theta) = 1$. Hence, it follows that the field determined by the harmonic with zero indices equals zero. Leaving $\alpha_{0,0}$ and $\beta_{0,0}$ out of consideration actually means that the values of the electric and magnetic potentials in the center of the receiving sphere are assumed to be equal to zero.

We take it that N spherical E-harmonics and the same quantity of H-harmonics are considered. They receive common numeration

$$n = 0,1,2,\ldots,N\text{-}1.$$

The $n1(n)$ and $n2(n)$ indices are calculated by:

$$n1(n) = \left\lfloor \sqrt{n+1} \right\rfloor, \quad n2(n) = n + 1 - n1(n) - n1^2(n) \tag{6.5}$$

In (6.5) $\lfloor x \rfloor$ denotes the integer part of x. The n dependence of indices $n1$ and $n2$ calculated by formula (6.5) for $N = 15$ is presented in Table 6.1.

Table 6.1. The relation between single-index and double-index notation of the spherical harmonics

n	0	1	2	3	4	5	6	7	8	9	10	11	12	13	14
$n1$	1	1	1	2	2	2	2	2	3	3	3	3	3	3	3
$n2$	-1	0	1	-2	-1	0	1	2	-3	-2	-1	0	1	2	3

In order to cover all $2n1_{max} + 1$ indices $n2$ corresponding to the maximum value of index $n1$ ($n1_{max}$), the total number of spherical harmonics N should be chosen in accordance with:

$$N = K^2 - 1 \tag{6.6}$$

where K denotes some integer value. For Table 6.1 above $K = 4$. If the total number of harmonics is large condition (6.6) may be disregarded. It is noteworthy that N is a number of harmonics of the same type, the overall number of spherical harmonics of both types thus being $2N$.

Let us now describe the numbering method, which is used for the plane wave sources. We will concern ourselves with two types of multipath channels. In the first case we operate on the premise that the sources are approximately evenly distributed over the sphere surface of a huge radius. With the number of the sources being large, this case represents the omnidirectional wireless channel.

It is assumed in the second case that the sources are located within some range
of angles $\theta_{min} \leq \theta_m \leq \theta_{max}$; $\varphi_{min} \leq \varphi_m \leq \varphi_{max}$, with either $\Delta\theta \ll \pi$ or $\Delta\varphi \ll \pi$, or
both inequalities satisfied. The case when one of the angles is fixed $\theta_{min} = \theta_{max} =$
$= \theta_0$ can be carried over here as well.

We assume that the number of the sources producing the waves of the same polarization is M, and use standard numbering for them

$$m = 0,1,2,\ldots,M\text{-}1.$$

In the first case, representing the omnidirectional multipath channel with uniform variation steps in the θ angle, the φ angle variation steps must not remain
fixed, decreasing as the equator is approached to offset the increase of the sphere
radius. For the omnidirectional wireless channel the following expressions can be
used for angular coordinate calculations. First compute the auxiliary indices $m_1(m)$
and $m_2(m)$.

$$m_1(m) = \left\lfloor \sqrt{m+1} \right\rfloor, \text{ for m} < M/2, \tag{6.7a}$$
$$m_1(m) = \left\lfloor \sqrt{M-m} \right\rfloor, \text{ for m} \geq M/2$$

$$m_2(m) = m+1-m_1(m)-m_1^2(m), \text{ for m} < M/2, \tag{6.7b}$$
$$m_2(m) = m - M + m_1(m) + m_1^2(m), \text{ for m} \geq M/2$$

With the help of the auxiliary indices, calculate the θ_m and φ_m angular coordinates of the sources by the formulae

$$\theta_m = \frac{\pi}{2} \cdot \frac{m_1(m)}{m_1\left(\dfrac{M-1}{2}\right)}, \text{ for m} < M/2, \tag{6.8a}$$

$$\theta_m = \pi - \frac{\pi}{2} \cdot \frac{m_1(m)}{m_1\left(\dfrac{M-1}{2}\right)}, \text{ for m} \geq M/2.$$

$$\varphi_m = 2\pi \frac{m_2(m)+m_1(m)}{1+2m_1(m)} \tag{6.8b}$$

For $M = 11$ sources, their angular coordinates computed by formulae (6.8) are
presented in Table 6.2.

Table 6.2. The angular coordinates of 11 wave sources in three-dimensional wireless
channel

m	0	1	2	3	4	5	6	7	8	9	10
$\theta_m^{\,0}$	45	45	45	90	90	90	90	90	135	135	135
$\varphi_m^{\,0}$	0	120	240	0	72	144	216	288	0	120	240

It is evident from Table 6.2 that the number of φ angle divisions in the equatorial plane is maximum and decreases as the sphere poles are approached ($\theta = 0$, $\theta = \pi$). Strictly speaking, the number of the points should vary proportionally to $\sin\theta$. Expressions (6.7) and (6.8) approximate the sinusoidal dependence with a linear one. To preserve the linearity of the variation, the number of sources M should be selected in accordance with

$$M = 2K^2 + 2K - 1 \tag{6.9}$$

Here K is an integer value. The value $M = 11$, taken in compiling Table 6.2 will be in consistency with $K = 2$.

In the second case, with a sector arrangement of the sources, the m dependence of their coordinates will be computed by:

$$\varphi_m = \varphi_{\min} + \mathrm{mod}(m, M_1)\frac{\Delta\varphi}{M_1 - 1} \tag{6.10a}$$

$$\theta_m = \theta_{\min} + \left\lfloor \frac{m}{M_1} \right\rfloor \frac{\Delta\theta}{M_2 - 1} \tag{6.10b}$$

In (6.10) $\mathrm{mod}(m, M_1)$ denotes the residue of dividing m by the integer M_1. The M_1 and M_2 values determine the number of the division points for the $\Delta\varphi$ and $\Delta\theta$ angle intervals respectively. The number of the sources for the waves of the same polarization is:

$$M = M_1 \cdot M_2 \tag{6.11}$$

The overall number of the (vertically and horizontally polarized) plane wave sources equals $2M$.

The numbering order selected, we now turn to considering the structure of the **H** channel matrix. This matrix links the amplitudes of plane waves (channel input) to the amplitudes of spherical harmonics (channel output). Let us collect the complex amplitudes of the input waves into vector **A**, and group the complex amplitudes of the output spherical harmonics into vector **B**. The relation between them is

$$\mathbf{B} = \mathbf{HA} \tag{6.12}$$

Here **H** is a $2N \times 2M$ matrix. Vector **A** has $2M$ components. Order them, first writing M complex amplitudes $\mathbf{A}^{\mathrm{v}}$ of the vertically polarized waves, and then M amplitudes $\mathbf{A}^{\mathrm{h}}$ of the horizontally polarized waves. Order the **B** vector, first writing N complex amplitudes of the E-waves, and then N amplitudes of the H-waves. In accordance with the above, the $2N \times 2M$ channel matrix **H** may be represented as an ensemble of four $N \times M$ sub-arrays.

$$\mathbf{H} = \begin{bmatrix} \mathbf{H}_E^v & \mathbf{H}_E^h \\ \mathbf{H}_H^v & \mathbf{H}_H^h \end{bmatrix} \qquad (6.13)$$

In each cellular matrix the v, h upper index indicates the polarization type of the source wave, while the lower index specifies the type of the spherical harmonic (E-type, H-type). The elements of the $\mathbf{H}^v{}_E$, $\mathbf{H}^v{}_H$, $\mathbf{H}^h{}_E$, $\mathbf{H}^h{}_H$ matrices are calculated with the help of formulae (6.4a, 6.4b, 6.4c and 6.4d) respectively. Since formulae a and d in (6.4) coincide, while formulae c and b differ by sign only, the $\mathbf{H}$ matrix (6.13) can be written as

$$\mathbf{H} = \begin{bmatrix} \mathbf{H}0 & -\mathbf{H}1 \\ \mathbf{H}1 & \mathbf{H}0 \end{bmatrix} \qquad (6.14)$$

The elements of matrices $\mathbf{H}0$ and $\mathbf{H}1$ are estimated from:

$$H0_{n,m} = i^{n1(n)-1} j_{n1(n)}(kr) e^{-in2(n)\varphi_m} \frac{d\left(PN_{n1}^{(n2(n))}(\cos\theta)\right)}{d\theta}\Bigg|_{\theta=\theta_m} \qquad (6.15a)$$

$$H1_{n,m} = i^{n1(n)} n2(n) j_{n1(n)}(kr) e^{-in2(n)\varphi_m} \frac{PN_{n1(n)}^{(n2(n))}(\cos\theta_m)}{\sin\theta_m} \qquad (6.15b)$$

Formulae (6.15) are employed in Chap. 7 for estimating the characteristics of the wireless channel with a spherical receiving area.

6.2 Translation of Plane Waves into Electric and Magnetic Potentials

In Sect. 6.2 and 6.3 we present channel estimation techniques other than those of Sect. 6.1. They permit checking if the calculation expressions of Sect. 6.1 are correct. Furthermore, different calculation methods must be applied if the receiving area is distinct from the sphere.

It is assumed in Sect. 6.2 that the receiver detects electric and magnetic potentials on the surface of a specified receiving area. That is, in contrast to the assumptions of Sect. 6.1, the receiver produces continuous functions U and V of the spatial coordinates on the surface of the receiving area rather than a discrete set of the complex amplitudes of spherical harmonics. However, in numerical evaluations the values of potentials will naturally be recorded at the discrete points of the receiving area. However, their number will be taken sufficiently large for the discrete functions to approximate the continuous potentials accurately enough.

The advantage of the approach adopted in Sect. 6.2 is in its applicability to both spherical areas and the receiving areas of other shapes. In what follows we consider three three-dimensional receiving areas: the ball-, the cylinder- and the parallelepiped-shaped. The ball-shaped area allows accuracy verification of the results obtained through use of the formulae of Sect. 6.1 and 6.2. The analysis of a cylinder-shaped area allows understanding of how channel limiting characteristics vary with an increase in one of the antenna's dimensions. The dimension size dependences of channel characteristics will become apparent during examination of the parallelepiped-shaped receiving area.

Consider the calculation formulae for the electric and magnetic potentials created by a set of plane wave sources on the surface of the receiving area. Regardless of the area shape they are based on representation of potentials by the sums (5.2, 5.3).

$$U_{l,n} = \sum_{n=0}^{N-1} \alpha_{n1(n),n2(n)} kr_l \, j_{n1(n)}(kr_l) PN_{n1(n)}^{(n2(n))}\left(\cos\theta_l\right) e^{in2(n)\varphi_l} \qquad (6.16a)$$

$$V_{l,n} = \sum_{n=0}^{N-1} \beta_{n1(n),n2(n)} kr_l \, j_{n1(n)}(kr_l) PN_{n1(n)}^{(n2(n))}\left(\cos\theta_l\right) e^{in2(n)\varphi_l} \qquad (6.16b)$$

Formulae (6.16) differ from (5.2) and (5.3) by the constant factor k^2. The constant factor does not change the matrix eigenvalues and is introduced to make the transition to dimensionless quantities. In (6.16) l denotes the current number of the point in the receiving area, which picks up field potentials. r_l, θ_l, φ_l stand for the spherical coordinates of the l-th point. The overall number of points is L, the numbering order and coordinate calculation procedures are considered hereafter. As in 6.1, m specifies the number of a plane wave. The angular coordinates of the wave sources are calculated by formulae (6.8) or (6.10). N signifies the number of spherical harmonics taken into account. Indices $n1$ and $n2$ are computed with regard to the current index n by formulae (6.5). $PN_{n1}^{(n2)}(\cos\theta)$ denotes the normalized associated Legendre function. Its values can be computed by formula (5.14) in terms of the normalizing factor (6.1). As before, $j_{n1}(kr_l)$ is the spherical Bessel function.

Expressions (6.16) demonstrate that it is convenient to present the channel matrix as a product of two arrays

$$\mathbf{H} = \mathbf{HU} \cdot \mathbf{HB} \qquad (6.17)$$

Here **HB** is the matrix representing conversion of the plane waves into the spherical harmonic coefficients α and β, while **HU** translates the coefficients α and β into the values of the potentials at discrete points. The sizes of matrices **HU** and **HB** are $2L \times 2N$ and $2N \times 2M$. Clearly, the size of matrix **H** is $2L \times 2M$.

The **HB** matrix is similar to the **H** matrix, evaluated by formula (6.15). Omit the $j(kr_l)$ factor from (6.15). It will be substituted into the **HU** matrix, which thus takes the form

$$\mathbf{HB} = \begin{bmatrix} \mathbf{HB0} & -\mathbf{HB1} \\ \mathbf{HB1} & \mathbf{HB0} \end{bmatrix} \qquad (6.18)$$

The elements of matrices **HB0** and **HB1** are computed by:

$$HB0_{n,m} = i^{n1(n)-1} e^{-in2(n)\varphi_m} \frac{d\left(PN_{n1}^{(n2(n))}(\cos\theta)\right)}{d\theta}\Bigg|_{\theta=\theta_m} \tag{6.19a}$$

$$HB1_{n,m} = i^{n1(n)} n2(n) e^{-in2(n)\varphi_{tm}} \frac{PN_{n1(n)}^{(n2(n))}(\cos\theta_m)}{\sin\theta_m} \tag{6.19b}$$

Formulae (6.16) permit derivation of the **HU** matrix. Considering that the coefficients α define only the electric potential and not the magnetic one, while the coefficient β, by contrast, determines only the magnetic potential, the **HU** array can be written as the cellular matrix

$$\mathbf{HU} = \begin{bmatrix} \mathbf{HU0} & \mathbf{0} \\ \mathbf{0} & \mathbf{HU0} \end{bmatrix} \tag{6.20}$$

Here **0** is a zero sub-array of size $L \times N$. As is obvious from (6.16) the elements of the **HU0** matrix are determined by:

$$HU0_{l,n} = kr_l j_{n1(n)}(kr_l) PN_{n1(n)}^{(n2(n))}(\cos\theta_l) e^{in2(n)\varphi_l} \tag{6.21}$$

Thus expressions (6.17-6.21) present the computational expressions for writing a channel matrix for picking up potentials at the discrete points of a receiving area surface. To be able to perform calculations, it is necessary to define a numbering order for the discrete points on the surface of the receiving area and their coordinate computation procedure. This issue is a prerequisite of the computation techniques and is solved differently for each type of the receiving areas under consideration.

Denote the number of the points on the surface of a receiving area, where the potentials are measured by L. l signifies a current (running) number

$$l = 0,1,2,\ldots, L\text{-}1. \tag{6.22}$$

For the receiving area of a spherical shape, use expressions (6.7 – 6.8) for numbering the wave sources and calculating the coordinates. Namely

$$l_1(l) = \left\lfloor \sqrt{l+1} \right\rfloor, \text{ for } l < L/2,$$
$$l_1(l) = \left\lfloor \sqrt{L-l} \right\rfloor, \text{ for } l \geq L/2 \tag{6.23a}$$

$$l_2(l) = l+1-l_1(l)-l_1^2(l), \text{ for } l < L/2,$$
$$l_2(l) = l-L+l_1(l)+l_1^2(l), \text{ for } l \geq L/2 \tag{6.23b}$$

With the help of the auxiliary indices, the angle coordinates θ_l and φ_l of the observation points are calculated by:

$$\theta_l = \frac{\pi}{2} \frac{l_1(l)}{l_1\left(\dfrac{L-1}{2}\right)} \,, \text{ for } l < L/2$$

$$\text{(6.24a)}$$

$$\theta_l = \pi - \frac{\pi}{2} \frac{l_1(l)}{l_1\left(\dfrac{L-1}{2}\right)} \,, \text{ for } l \ge L/2$$

$$\varphi_l = 2\pi \frac{l_2(l) + l_1(l)}{1 + 2l_1(l)} \tag{6.24b}$$

For a sphere of radius r the radial coordinate remains unchanged

$$r_l = r \tag{6.24c}$$

There is good reason to choose L in accordance with equality (6.9)

$$L = 2K^2 + 2K - 1 \tag{6.25}$$

Here K is an integer value.

Consider the calculation expressions for the coordinates of a discrete point of a cylinder-shaped receiving area. Assume that the axis of symmetry of the receiving area coincides with the z axis. The cylinder height will be denoted by a, and b signifies the cylinder base radius. Assume also that the approximate size Δ_{ap} of the cell, into which the cylinder surface will be broken up, is specified. The true value of the height step Δ and the number of points L on the surface of the cylinder is determined by:

$$\Delta = \frac{a}{\left\lfloor \dfrac{a}{\Delta_{ap}} \right\rfloor + \delta\left(\left\lfloor \dfrac{a}{\Delta_{ap}} \right\rfloor\right)} \tag{6.26a}$$

$$L_b = \left\lfloor \frac{b}{\Delta} \right\rfloor - \delta\left(b - \left\lfloor \frac{b}{\Delta} \right\rfloor\right) \tag{6.26b}$$

$$L_0 = \frac{a}{\Delta} + 2L_b + 1 \tag{6.26c}$$

$$L_c = \left\lfloor \frac{2\pi b}{\Delta} \right\rfloor + \delta\left(\left\lfloor \frac{2\pi b}{\Delta} \right\rfloor\right) \tag{6.26d}$$

$$L = L_0 L_c \tag{6.26e}$$

In expressions (6.26) the brackets $\lfloor \ \rfloor$ denote the integer part, δ stands for the delta-function (it is unity, if the argument is zero, otherwise it is zero). L_c denotes the number of points within the angle φ, L_b signifies the number of the break-up points in the base radius, L_0 is the total number of the points in the cylinder generating line.

The number of a point on the cylinder surface varies in a conventional way (6.22), and the cylindrical coordinates of a point are calculated by:

$$
rc_l = \begin{cases} \left(b - \left(L_b - \left\lfloor \dfrac{l}{L_c} \right\rfloor \right) \right)\Delta \ \dots\dots\dots \ for \dots \left\lfloor \dfrac{l}{L_c} \right\rfloor \leq L_b \\[3ex] b \ \dots\dots\dots\dots\dots\dots\dots \ for \dots L_b < \left\lfloor \dfrac{l}{L_c} \right\rfloor < L_0 - L_b - 1 \\[3ex] \left(b - \left(L_b + 1 + \left\lfloor \dfrac{l}{L_c} \right\rfloor - L_0 \right)\Delta \right) \ \dots\dots \ for \dots \left\lfloor \dfrac{l}{L_c} \right\rfloor \geq L_0 - L_b - 1 \end{cases}
$$

(6.27a)

$$
\varphi c_l = \mathrm{mod}(l, L_c) \frac{2\pi}{L_c}
$$

(6.27b)

$$
zc_l = \begin{cases} \dfrac{a}{2} \ \dots\dots\dots\dots \ for \dots \left\lfloor \dfrac{l}{L_c} \right\rfloor \leq L_b \\[3ex] \dfrac{a}{2} - \left(\left\lfloor \dfrac{l}{L_c} \right\rfloor - L_b \right)\Delta \ \dots\dots \ for \dots L_b < \left\lfloor \dfrac{l}{L_c} \right\rfloor < L_0 - L_b - 1 \\[3ex] -\dfrac{a}{2} \ \dots\dots\dots\dots \ for \dots \left\lfloor \dfrac{l}{L_c} \right\rfloor \geq L_0 - L_b - 1 \end{cases}
$$

(6.27c)

For transition from the cylindrical coordinates rc_l, φc_l and zc_l to the spherical r_l, θ_l and φ_l the usual formulae are used:

$$
\eta = \sqrt{rc_l^2 + zc_l^2} \ , \quad \sin\theta_l = \frac{rc_l}{\eta} , \quad \varphi_l = \varphi c_l
$$

(6.28)

The parallelepiped-shaped receiving area will be defined by the dimensions a, b and c along the axes x, y and z respectively. It is assumed also that the approximate size Δ_{ap} of a cell, into which the surface of the parallelepiped will be broken up, is specified. The amount of the points on the parallelepiped surface L and their coordinates are determined by the relationships:

$$
L_x = \left\lfloor \frac{a}{\Delta_{ap}} \right\rfloor + 1 , \quad L_y = \left\lfloor \frac{b}{\Delta_{ap}} \right\rfloor + 1 , \quad L_z = \left\lfloor \frac{c}{\Delta_{ap}} \right\rfloor + 1
$$

(6.29a)

$$L_{xy} = 2(L_x + 1)(L_y + 1), \quad L_{xz} = 2(L_x + 1)(L_z - 1), \quad L_{yz} = 2(L_y - 1)(L_z - 1) \tag{6.29b}$$

$$L = L_{xy} + L_{xz} + L_{yz} \tag{6.29c}$$

$$x_l = \begin{cases} -\dfrac{a}{2} + \dfrac{a}{L_x} \mathrm{mod}\left(\left\lfloor \dfrac{l}{2} \right\rfloor, L_x + 1 \right) \dots\dots\dots for\dots l < L_{xy} + L_{xz} \\[4mm] (-1)^l \dfrac{a}{2} \dots\dots\dots\dots\dots\dots\dots\dots\dots\dots\dots for\dots l \ge L_{xy} + L_{xz} \end{cases} \tag{6.30a}$$

$$y_l = \begin{cases} -\dfrac{b}{2} + \dfrac{b}{L_y} \dfrac{\left\lfloor \dfrac{l}{2} \right\rfloor - \mathrm{mod}\left(\left\lfloor \dfrac{l}{2} \right\rfloor, L_x + 1 \right)}{L_x + 1} \dots\dots\dots for\dots l < L_{xy} \\[8mm] (-1)^l \dfrac{b}{2} \dots\dots\dots\dots\dots\dots\dots\dots\dots\dots\dots for\dots L_{xy} \le l < L_{xy} + L_{xz} \\[4mm] -\dfrac{b}{2} + \dfrac{b}{L_y}\left(1 + \mathrm{mod}\left(\left\lfloor \dfrac{l - L_{xy} - L_{xz}}{2} \right\rfloor, L_y - 1 \right) \right) \dots for\dots l \ge L_{xy} + L_{xz} \end{cases} \tag{6.30b}$$

$$z_l = \begin{cases} (-1)^l \dfrac{c}{2} \dots\dots\dots\dots\dots\dots\dots\dots\dots\dots\dots\dots\dots\dots for\dots l < L_{xy} \\[8mm] -\dfrac{c}{2} + \dfrac{c}{L_z}\left(1 + \dfrac{\left\lfloor \dfrac{l - L_{xy}}{2} \right\rfloor - \mathrm{mod}\left(\left\lfloor \dfrac{l - L_{xy}}{2} \right\rfloor, L_x + 1 \right)}{L_x + 1} \right) \dots\dots for\dots L_{xy} \le l < L_{xy} + L_{xz} \\[10mm] -\dfrac{c}{2} + \dfrac{c}{L_z}\left(1 + \dfrac{\left\lfloor \dfrac{l - L_{xy} - L_{xz}}{2} \right\rfloor - \mathrm{mod}\left(\left\lfloor \dfrac{l - L_{xy} - L_{xz}}{2} \right\rfloor, L_y - 1 \right)}{L_y - 1} \right) for\dots l \ge L_{xy} + L_{xz} \end{cases} \tag{6.30c}$$

The spherical coordinates r_l, θ_l and φ_l of the discrete points on the surface of the parallelepiped are calculated from the Cartesian coordinates x_l, y_l and z_l

$$r_l = \sqrt{x_l^2 + y_l^2 + z_l^2}, \quad \cos\theta_l = \frac{z_l}{r_l}, \quad tg\varphi_l = \frac{y_l}{x_l} \tag{6.31}$$

Other receiving areas of different shapes can be considered in a similar manner.

6.3 Calculation of the Channel Matrix Elements in Terms of the Source Wave to Field Components Translation

In this section we assume that the receiver analyzes the electric and magnetic fields at discrete points on the surface of the receiving area. It means that in writing the **HF** channel matrix we do not refer to the concept of potential, but assume instead that the **HF** matrix establishes the relation between the **A** vector (the complex amplitudes of vertically and horizontally polarized plane waves) and vector **F** (the field complex amplitudes at the grid nodes on the surface of the receiving area).

$$\mathbf{F} = \mathbf{HF} \cdot \mathbf{A} \tag{6.32}$$

First, we proceed from the assumption that the receiver picks up the Cartesian components of the electric and magnetic fields at discrete points on the surface of the receiving area. That is the **F** vector contains the complex amplitudes of fields E_x, E_y, E_z, H_x, H_y, H_z at the nodal points of the grid on the surface of the receiving area. Then, we include only the tangent components of the E- and H fields into the **F** vector.

To derive the estimation relationships for the elements of the **HF** matrix we should write the field (determined by the angles θ_t and φ_t) of an arbitrarily directed plane wave. For the vertically polarized wave:

$$E_x = E \cos\theta_t \cos\varphi_t e^{i(\mathbf{kr})}, \quad E_y = E \cos\theta_t \sin\varphi_t e^{i(\mathbf{kr})}$$
$$E_z = -E \sin\theta_t e^{i(\mathbf{kr})}, \quad H_x = H \sin\varphi_t e^{i(\mathbf{kr})} \tag{6.33a}$$
$$H_y = -H \cos\varphi_t e^{i(\mathbf{kr})}, \quad H_z = 0$$

For the horizontally polarized wave:

$$E_x = -E \sin\varphi_t e^{i(\mathbf{kr})}, \quad E_y = E \cos\varphi_t e^{i(\mathbf{kr})}$$
$$E_z = 0, \quad H_x = H \cos\theta_t \cos\varphi_t e^{i(\mathbf{kr})} \tag{6.33b}$$
$$H_y = H \cos\theta_t \sin\varphi_t e^{i(\mathbf{kr})}, \quad H_z = -H \sin\theta_t e^{i(\mathbf{kr})}$$

In (6.33) the amplitudes of the electric and magnetic fields are linked by a factor representing the characteristic wave impedance of the medium.

$$E = \sqrt{\frac{\mu}{\varepsilon}} H$$

In what follows, it is assumed that $E = 1$, $H = 1$, as in (6.4). Use of the equalities $E = 1$, $H = 1$ implies that picking up the magnetic field of a plane wave permits extraction of exactly the same amount of information as picking up its electric field. $\mathbf{kr}$ in (6.33) denotes a scalar product of the wave vector (the vector of size $\mathbf{k}$, whose direction is determined by the angles θ_t and φ_t) and the $\mathbf{r}$ vector, directed from the coordinate origin to the observation point. Having written the $\mathbf{k}$ and $\mathbf{r}$ vectors in terms of their Cartesian components

$$\mathbf{k} = k \sin \theta_t \cos \varphi_t \mathbf{e}_x + k \sin \theta_t \sin \varphi_t \mathbf{e}_y + k \cos \theta_t \mathbf{e}_z$$

$$\mathbf{r} = x \mathbf{e}_x + y \mathbf{e}_y + z \mathbf{e}_z$$

we obtain the expression for the scalar product

$$(\mathbf{kr}) = kx \sin \theta_t \cos \varphi_t + ky \sin \theta_t \sin \varphi_t + kz \cos \theta_t \qquad (6.34a)$$

Expression (6.34a) can be written differently, in terms of the spherical coordinates (r, θ, φ) of the observation point

$$(\mathbf{kr}) = kr \sin \theta \cos \varphi \sin \theta_t \cos \varphi_t + kr \sin \theta \sin \varphi \sin \theta_t \sin \varphi_t + kr \cos \theta \cos \theta_t \qquad (6.34b)$$

Written in terms of the cylindrical coordinates of the observation point (rc, φc, zc) the expression takes the form

$$(\mathbf{kr}) = krc \cos \varphi \sin \theta_t \cos \varphi_t + krc \sin \varphi \sin \theta_t \sin \varphi_t + kz \cos \theta_t \qquad (6.34c)$$

Expressions (6.34a, 6.34b, 6.34c) are convenient for use in analysis of parallelepiped-shaped receiving areas, spherical receiving areas and cylindrical receiving areas respectively.

With the assumption that the components of the $\mathbf{F}$ vector are arranged in the following order E_x, E_y, E_z, H_x, H_y, H_z, let us write the $\mathbf{HF}$ channel matrix as that consisting of twelve $L \times M$ sub-arrays. L denotes the number of the field pick-up points; M stands for the number of the plane wave sources.

$$\mathbf{HF} = \begin{bmatrix} \mathbf{HE}_x^v & \mathbf{HE}_x^h \\ \mathbf{HE}_y^v & \mathbf{HE}_y^h \\ \mathbf{HE}_z^v & \mathbf{HE}_z^h \\ \mathbf{HH}_x^v & \mathbf{HH}_x^h \\ \mathbf{HH}_y^v & \mathbf{HH}_y^h \\ \mathbf{HH}_z^v & \mathbf{HH}_z^h \end{bmatrix} \qquad (6.35)$$

In the cellular matrices of (6.35), the upper index (v, h) denotes the polarization of the incident plane wave, the subscript (x, y, z) signifies the projection axis for the picked up field, the second literal (E, H) represents the type of the registered field.

The twelve formulae presented above (6.33) are the expressions for calculation of the cellular matrix elements of (6.35). In numerical evaluation of the $HF_{l,m}$ elements of the matrices we will adhere to the numbering order m adopted in Sect. 6.1 for the plane wave sources and the numbering order l adopted in Sect. 6.2 for the grid nodal points on the surface of the receiving area. It is known from electrodynamics that a full field, in some limited region of space free from sources, is determined by specifying the field tangent components along the region boundary. Thus, we can diminish the size of the **HF** matrix in (6.32), under the assumption that the receiver picks up the tangent components rather than the Cartesian components of the field. In such a case the **HF** matrix takes the form as in (6.35), but will be of a somewhat smaller size.

$$\mathbf{HF} = \begin{bmatrix} HE_1^v & HE_1^h \\ HE_2^v & HE_2^h \\ HH_1^v & HH_1^h \\ HH_2^v & HH_2^h \end{bmatrix} \tag{6.36}$$

In order to compute the sub-array elements of matrix (6.36) there is good reason to express the tangent field components in terms of Cartesian components. This relationship is different for each of the considered areas. For example, for the spherical area the tangent field components $E_\theta, E_\varphi, H_\theta, H_\varphi$ in terms of expressions (5.51) and (6.33) are equal with vertical polarization of the incident plane wave:

$$E_\theta = E(\cos\theta_t \cos\varphi_t \cos\theta \cos\varphi + \cos\theta_t \sin\varphi_t \cos\theta \sin\varphi + \sin\theta_t \sin\theta)e^{i(\mathbf{kr})}$$

$$E_\varphi = E(-\cos\theta_t \cos\varphi_t \sin\varphi + \cos\theta_t \sin\varphi_t \cos\varphi)e^{i(\mathbf{kr})}$$

$$H_\theta = H(\sin\varphi_t \cos\theta \cos\varphi - \cos\varphi_t \cos\theta \sin\varphi)e^{i(\mathbf{kr})} \tag{6.37a}$$

$$H_\varphi = H(-\sin\varphi_t \sin\varphi - \cos\varphi_t \cos\varphi)e^{i(\mathbf{kr})}$$

For a horizontally polarized wave:

$$E_\theta = E(-\sin\varphi_t \cos\theta \cos\varphi + \cos\varphi_t \cos\theta \sin\varphi)e^{i(\mathbf{kr})}$$

$$E_\varphi = E(\sin\varphi_t \sin\varphi + \cos\varphi_t \cos\varphi)e^{i(\mathbf{kr})}$$

$$H_\theta = H(\cos\theta_t \cos\varphi_t \cos\theta \cos\varphi + \cos\theta_t \sin\varphi_t \cos\theta \sin\varphi + \sin\theta_t \sin\theta)e^{i(\mathbf{kr})} \tag{6.37b}$$

$$H_\varphi = H(-\cos\theta_t \cos\varphi_t \sin\varphi + \cos\theta_t \sin\varphi_t \cos\varphi)e^{i(\mathbf{kr})}$$

The receiving areas of other shapes may be considered in a similar way.

6.4 Summary

Three distinct ways of channel matrix presentation for the three-dimensional wireless channel have been covered in Chap. 6.

1. In Sect. 6.1 the receiver is viewed as extracting information from the electromagnetic field through analysis of the complex amplitudes of spherical harmonics.
2. The matrix representation in Sect. 6.2 is based on the assumption that the receiver picks up the values of the electric and magnetic potentials at discrete points on the surface of the receiving area.
3. In the perspective of Sect. 6.3 the discrete points over the surface of the receiving area are regarded as pick-up points for the projections of the electric and magnetic fields.

7 Limit Capacity of the Three-Dimensional Wireless Channel with Potential Antennas

This chapter sets out the results of numerical evaluation of the maximum attainable channel capacity with a preassigned receiving area in the 3-D space. We proceed from the assumption that the maximum channel capacity can be achieved through employment of some speculative potential receiving antenna. The term potential antenna owes its origin to two factors. First, with use of such antenna the limit (potential) performance characteristics of the channel are attained. Second, its estimation is based on the concept of the electric and magnetic potentials introduced in Chap. 5.

In the estimation methods and the nature of the obtained results Chap. 7 resembles Sect. 4.3. The dissimilarity between the two is found in that here we concern ourselves with investigating the three-dimensional channel as differentiated from the 2-D channel tackled previously. Besides, in order not to make the chapter seem unduly dilated and verbose, many of the analysis results had to be dropped out of the discussion. For example, here we do not consider interference noise, limiting ourselves to presenting the analysis results for thermal noise only. Introduction of interference noise in lieu of thermal noise leads to a rise in the optimal number of spatial subchannels and augments the limit capacity. In the four sections of Chap. 7 we examine four types of receiving areas: the sphere, the cylinder, the parallelepiped and the rectilinear segment.

7.1 Limiting Characteristics of the Multipath Wireless Channel with a Spherical Receiving Area

The 3-D wireless channel with a spherical-shaped receiving area will be our initial concern. First, as with the 2-D channel, let us perform an approximate estimation of the wireless channel limit capacity. Such estimations are based on the properties of the spherical Bessel functions $j_n(x)$, included in expressions (6.15) for calculation of the channel matrix elements.

For an electrically small receiving area $kr \ll 1$, only the elements of the biggest module need to be preserved in the **H** matrix (6.14). These are the elements with the index $n = 1$, of the order kr, as is evident from (5.12). We are reminded that in writing the **H** matrix the elements with zero index $n = 0$, representing constant potentials and zero fields, have been omitted.

Three associated Legendre functions $P_1^{(m)}(\theta)$ with the index m equal to 0, -1 and 1 correspond to index $n = 1$. It means that with an electrically small receiving area only 6 approximately equal summands need to be left within the **H** matrix, 3 for each type of the waves (E-waves and H-waves), allowing for 6 spatial subchannels.

$$N_{ap} = 6 \qquad (7.1)$$

Thus, with the omnidirectional 3-D wireless channel no spatial separation is needed to set up a communication system with six spatial subchannels. The results of numerical evaluation presented below are evidence in favor of the conclusion that in the omnidirectional wireless channel the optimal number of spatial subchannels always proves to be no less than 6.

It is possible to provide the following physical explanation for this fact. The electromagnetic field at an arbitrary point of space is characterized by six scalar quantities – three projections of the E vector and three projections of the H vector. In the omnidirectional channel neither type of projection can be favored over the other, since they are equipollent.

With an electrically large spherical receiving area $kr \gg 1$, to perform an approximate estimation of the number of spatial subchannels we use the property of the spherical Bessel functions $j_n(x)$, illustrated by the graph in Fig.7.1.

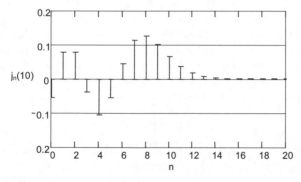

Fig. 7.1. The n index dependence of the values of the $j_n(x)$ spherical Bessel functions with the fixed argument $x = 10$

The $j_n(x)$ values quickly decrease as n grows, if the function index exceeds the value of the argument. The plot in Fig. 7.1, which portrays the dependence of $j_n(x)$ with $x = 10$, pictorially summarizes this property. This property, being similar to the property of the $J_n(x)$ Bessel function of the first kind, permits us, in writing the **H** channel matrix elements, to limit the size of the n index by the value:

$$n_{max} = kr \qquad (7.2)$$

Taking into account that for every value of the index n there are $2n + 1$ values of the index m, we may write

$$\sum_{n=1}^{n_{max}} (2n + 1) = (n_{max} + 1)^2 - 1$$

Hence, neglecting the unities, in terms of (7.2) and allowing for two types of the waves, we obtain the approximate expression for estimating the number of spatial subchannels

$$N_{ap} \approx 2(kr)^2 \tag{7.3}$$

or, expressing k in terms of the wavelength λ_w, we have

$$N_{ap} \approx 8\pi^2 \frac{r^2}{\lambda_w^2} \tag{7.4}$$

Estimation of the number of spatial subchannels permits rough evaluation of the limit capacity per unit bandwidth for the wireless channel

$$C_{ap} \approx N_{ap} \log\left(1 + \frac{SNR}{N_{ap}}\right) \tag{7.5}$$

The (7.1 - 7.5) estimates are rather rough and in pursuance of more accurate evaluation we now turn to the results of numerical analysis.

The plots for the SNR dependence of the optimal number of subchannels and the limit capacity for a spherical receiving area are presented in Fig. 7.2 – 7.6. Fig. 7.2 depicts the results of investigating a small-size spherical receiving area ($r = 0.04\lambda_w$) in the 3-D omnidirectional channel. The computations have been performed by the formulae of Chap. 3, with the allowance for presence of thermal noise only. The **H** channel matrix has been calculated on the basis of the spherical harmonics analysis (formulae 6.4). In the calculations the number of the sources of waves of each polarization type was taken to be 111.

The plot in Fig. 7.2a substantiates our conclusion that with a small-size receiving area the optimal number of subchannels in the 3-D omnidirectional channel is 6. The numerical analysis establishes that the number of subchannels turns out in excess of 6 only with high SNR, over 35dB in this example. In such a case the number of spatial subchannels shows a rise from 6 to 16. Such a rise is accounted for by presence of spherical harmonics with index $n = 2$. For $n = 2$ there are 5 values of the m index for every type of the waves (E-waves and H-waves).

A big number of spatial subchannels with a small-sized receiving area actually means employment of a superdirective receiver antenna. Therefore, a limited SNR is the most critical deterrent to the use of superdirectivity.

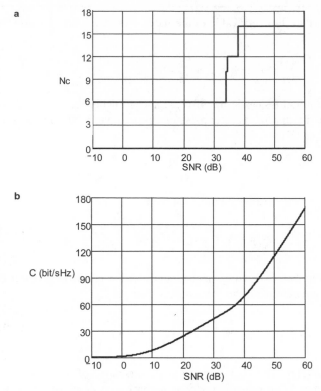

Fig. 7.2. SNR dependence of the characteristics of a 3-D omnidirectional wireless channel where **a** represents the optimal number of spatial subchannels and **b** represents the maximum capacity. The receiving area is a small-radius sphere ($r = 0.04\lambda_w$). The number of the sources of the waves of each polarization equals 111, the number of the spherical harmonics taken into account is 35 for each type of the waves.

Figure 7.3 illustrates how the characteristics of the omnidirectional channel vary with the radius variations of the spherical receiving area. The presented results account for three radius values $r = 0.04\ \lambda_w$, $r = 0.2\lambda_w$ and $r = \lambda_w$. During computations the number of wave sources was taken equal to 111 for each type of wave polarization, with the number of recorded spherical harmonics being 80.

It is evident from comparison of the plots in Fig. 7.3a and 7.3b for $r = 0.04\ \lambda_w$ and $r = 0.2\lambda_w$ that the spherical receiving area may be viewed as electrically small with the radius $r \leq 0.2\ \lambda_w$. The plots for $r = 0.04\ \lambda_w$ and $r = 0.2\lambda_w$ practically coincide. The optimal number of spatial subchannels is 6 in both cases.

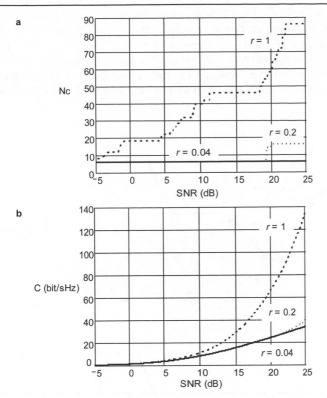

Fig. 7.3. SNR dependence of the characteristics of the 3-D omnidirectional wireless channel with spherical receiving areas. The sphere radius r is expressed in the wavelength units. The number of the wave sources is 111 for the waves of each polarization type, the number of the picked up spherical harmonics taken into account in the computations is 80. Chart **a** represents the optimal number of spatial subchannels. Chart **b** represents the maximum capacity.

The appearance of the plot for $r = \lambda_w$ in Fig. 7.3a is noticeably distinct from the plots for $r = 0.04\lambda_w$ and $r = 0.2\lambda_w$. An almost continuous rise in the optimal number of spatial subchannels with a gain in SNR can be observed here. The plot for $r = \lambda_w$ in Fig. 7.3b shows that with SNR = 15dB the limit capacity equals 30 bps/Hz, and with SNR = 25dB it becomes 130 bps/Hz. The plot in Fig. 7.3a shows that 46 and 86 spatial subchannels are needed to secure such enormous transfer rates in the receiving area of the radius $r = \lambda_w$. Therefore, in the three-dimensional wireless channel it is not so much big spatial dimensions as high SNR values that ensure high data transfer rates.

Figure 7.4 presents a comparison of the results of various analysis techniques used in estimation of the characteristics of the omnidirectional channel with a spherical receiving area. Curve 1 in both Fig. 7.4a and 7.4b has been calculated as in Fig. 7.3, in terms of spherical harmonics. For Curve 2, the estimation of the **H** channel matrix was performed with the assumption that the receiver picks up the tangent components of the electric and magnetic fields on the surface of the sphere. In other words, the **H** matrix has been calculated by formulae (6.38). The computation has been carried out with the assumption that the number of sources of waves of every polarization type equals 111, and the number of pick-up points for the tangent field components is 83. The radius of the spherical area equals the wavelength ($r = \lambda_w$).

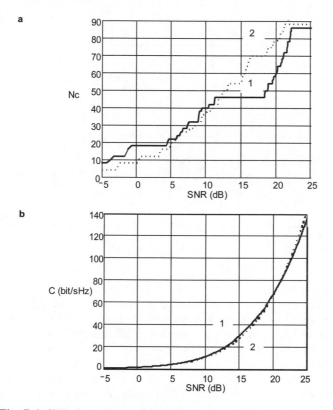

Fig. 7.4. SNR dependence of the characteristics of the 3-D omnidirectional wireless channel with a spherical receiving area ($r = \lambda_w$). The number of sources of waves of each polarization type is 111. Curves 1 represent channel estimation by spherical harmonics. The number of the spherical harmonics taken into account during estimation is 83 for each wave type. Curves 2 represent estimation of the characteristics by the tangent field components of the spherical surface. The number of discrete points on the sphere surface is 83. Chart **a** represents the optimal number of spatial subchannels. Chart **b** represents the maximum capacity.

A comparison of Curves 1 and 2 in Fig. 7.4b shows that various estimation techniques yield similar though not identical limit capacity values. The difference between the obtained results is probably accounted for by errors in numerical

computations. These errors cause certain discrepancies in the eigenvalues. In Fig.
7.4a the disaccord between Curves 1 and 2 is more pronounced, especially for
small and large SNR values.

In the course of plotting the graphs in Fig. 7.5 it was assumed that all the wave
sources are located in the horizontal plane ($\theta = \pi/2$). We will refer to this channel
as the omnidirectional azimuth channel as distinct from the omnidirectional 3-D
channel considered in Fig. 7.2-7.4. An arrangement of sources in a single horizontal plane makes such a channel similar to the 2-D wireless channel.

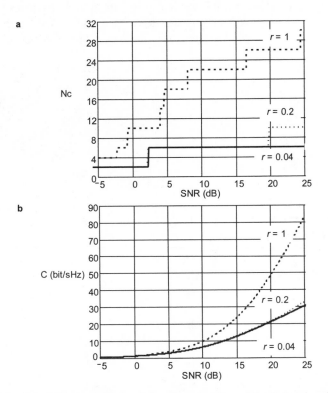

Fig. 7.5. SNR dependence of the characteristics of the 3-D azimuth wireless channel with
spherical receiving areas where **a** represents the optimal number of spatial subchannels **b**
represents the maximum capacity. The sphere radius r is given in wavelength units. The
number of sources of waves of each polarization type is 111, the number of the spherical
harmonics taken into account during computations is 80 for each wave type.

Figure 7.5 shows how the characteristics of the azimuth channel vary with radius variations of the spherical receiving area. The presented results account for
three values of the radius ($r = 0.04\ \lambda_w$, $r = 0.2\lambda_w$ and $r = \lambda_w$). The calculation has
been performed by the formulae of Chap. 3, with the allowance for presence of
thermal noise only. The **H** channel matrix computation is based on the spherical
harmonics analysis (formulae 6.4). In the course of calculations, the number of the

sources of the waves of each polarization type was taken to be 111, the number of picked-up spherical harmonics was 80.

A comparison between the azimuth and omnidirectional 3-D wireless channel (Fig. 7.5 and Fig. 7.3) reveals that a dimensional limitation imposed on spatial arrangement of the wave sources leads to a decrease in the optimal number of spatial subchannels and a drop in the attainable limit capacity of the channel. The plots for r =0.04 λ_w and r = 0.2λ_w in Fig. 7.5a demonstrate that with small-sized receiving areas, the azimuth channel becomes similar to the 2-D omnidirectional wireless channel discussed in Sect. 4.3. The optimal number of spatial subchannels in it is 2 for a small SNR value. However, this similarity is limited to SNR values < 5dB only. With greater SNRs the azimuth channel takes on a resemblance to the 3-D omnidirectional wireless channel. The optimal number of spatial subchannels in it becomes equal to 6.

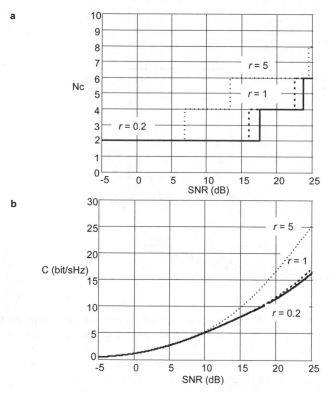

Fig. 7.6. SNR dependence of the characteristics of the 3-D sector channel with spherical receiving areas ($0 \leq \theta \leq .30^0$, $-30^0 \leq \varphi \leq 30^0$). The r sphere radius is given in wavelengths. The number of sources of waves of each polarization type is 200, the number of the spherical harmonics taken into account during computations is 80 for each type of the waves. Chart **a** represents the optimal number of spatial subchannels Chart **b** represents the maximum capacity.

Figure 7.6 shows the results of analyzing a sector arrangement of the sources with a spherical receiving area. It was assumed that 200 wave sources are located in the sector of the 30 degree angle θ and 60 degree azimuth. The angular separation between the sources is 3 degrees (10 different θ values and 20 φ values).

The results are presented for three radius values of the spherical receiving area ($r = 0.2\ \lambda_w$, $r = \lambda_w$ and $r = 5\lambda_w$). The calculation has been performed by the formulae of Chap. 3, with the allowance for presence of thermal noise only. The $\mathbf{H}$ channel matrix has been computed based on the spherical harmonics analysis (formulae 6.4). The number of the spherical harmonics taken into account during estimation was 80.

A comparison of the plots in Fig. 7.6 and Fig. 7.3 shows that a limitation imposed on spatial arrangement of the sources results in a decrease of the optimal number of subchannels and the attainable limit capacity.

7.2 Limiting Characteristics of the Three-Dimensional Wireless Channel with a Cylinder-Shaped Receiving Area

Let us consider a three-dimensional wireless channel with a cylinder-shaped receiving area. It is assumed that the cylinder axis coincides with the z-axis of the coordinate system. In estimation of the channel matrix we assume that Cartesian components of the electric and magnetic fields are picked up at the discrete points on the cylinder surface. That is, we make use of the relationships of Sect. 6.3.

Figures 7.7–7.9 depict the results of analyzing the three-dimensional omnidirectional channel. The θ and φ angular coordinates of the wave sources are calculated by formulae 6.10, in the computations the number of wave sources was taken to be equal to 59. Cylinders of various sizes and different height-to-base radius ratios have been considered.

Figure 7.7 presents the SNR dependence plots for the optimal number of spatial subchannels and the limit capacity of the cylinder, whose height a is equal to the base diameter ($a = 2b$). An electrically small cylinder has been under consideration ($a = 0.2\lambda_w$, $b = 0.1\lambda_w$), as well as a cylinder whose size is commensurate with the wavelength ($a = 2\lambda_w$, $b = \lambda_w$). It has been further assumed that the fields are picked up at 375 points on the cylinder surface.

It is instructive the compare the characteristics represented in Fig. 7.7 with the estimation results for the spherical receiving areas depicted in Fig. 7.3. The results for the cylinder ($a = 2\lambda_w$, $b = \lambda_w$) seem to be close to those for the sphere of radius $r = \lambda_w$, while the results for the cylinder ($a = 0.2\lambda_w$, $b = 0.1\lambda_w$) agree with the results obtained for the electrically small sphere. It might be well to point out that in Fig.7.7a for $a = 2\lambda_w$, $b = \lambda_w$ the variation in the number of subchannels is less abrupt than in Fig. 7.3a for $r = \lambda_w$.

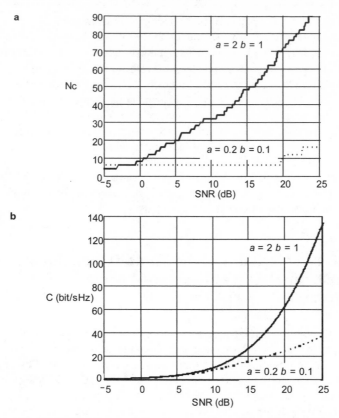

Fig. 7.7. SNR dependence of the characteristics of the 3-D omnidirectional channel with cylinder-shaped receiving areas. SNR dependence of the characteristics of the 3-D omnidirectional channel with cylinder-shaped receiving areas. The cylinder height a and the base radius b are given in wavelength units. The number of the sources of each wave polarization type is 59, the number of the discrete points on the cylinder surface, where the electric and magnetic fields are picked up equals 375. Chart **a** represents the optimal number of spatial subchannels. Chart **b** represents the maximum capacity.

Figure 7.8 presents the characteristics of the 3-D omnidirectional channel with a disk-shaped cylinder receiving area. We have examined small-height cylinders $(a = 0.1\lambda_w, b = \lambda_w)$ and $(a = 0.01\lambda_w, b = 0.1\lambda_w)$. The number of sources of each wave polarization type is 59, the number of field recording points is 500.

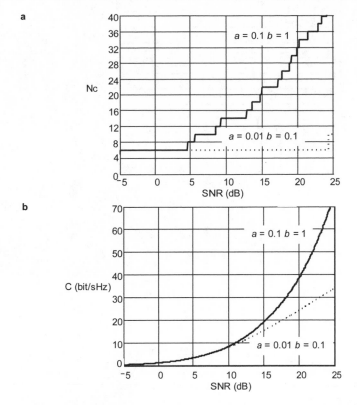

Fig. 7.8. SNR dependence of the characteristics of the 3-D omnidirectional channel with the disk-like cylinder receiving areas. The cylinder height a and the base radius b are given in wavelengths. The number of sources of every wave polarization type is 59, the number of discrete electric and magnetic field pick-up points on the cylinder surface is 500. Chart **a** represents the optimal number of spatial subchannels. Chart **b** represents the maximum capacity.

A comparison of the plots in Fig. 7.8 with those in Fig. 7.7 and 7.3 demonstrates that a transition to the disk-shaped receiving area does not cause much change in the characteristics of the omnidirectional wireless channel. An electrically small disk produces 6 spatial subchannels. For a disk size commensurate with the wavelength, the number of spatial subchannels increases monotonically with a rise in SNR. The diminished number of subchannels and a drop in the channel capacity in Fig. 7.8 in comparison with those in Fig. 7.7 for $r = \lambda_w$ is accounted for by the diminished size of the receiving area.

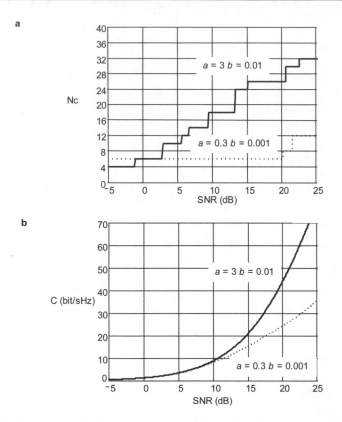

Fig. 7.9. SNR dependence of the characteristics of the 3-D omnidirectional channel for rod-like cylinder receiving areas. The cylinder height a and the base radius b are expressed in wavelengths. The number of horizontally and vertically polarized wave sources is 59 each, the number of discrete points on the cylinder surface picking up the electric and magnetic fields is 61. Chart **a** represents the optimal number of spatial subchannels. Chart **b** represents the maximum capacity.

Presented in Fig. 7.9 are the characteristics of the 3-D omnidirectional channel with a rod-like cylinder receiving area. Cylinders extended along the z-axis have been considered ($a = 3\lambda_w$, $b = 0.01\lambda_w$) and ($a = 0.3\lambda_w$, $b = 0.001\lambda_w$). The number of horizontally and vertically polarized wave sources is 59 each; the number of field pick-up points is 61. The discrete points are located along one of the cylinder elements. It is pertinent to note that the rod-like receiving area has nothing to do with the rod antenna, but is rather comparable to an antenna array. The number of antenna array elements should be taken to be six times the number of discrete points in the receiving area. Three components of the electric and three components of the magnetic field are recorded at each discrete point.

The conclusions drawn regarding the omnidirectional channel with the rod-like receiving area are similar to the inference for the disk receiving area. A comparison of the plots in Fig. 7.9 with the graphs in Fig. 7.7 and 7.3 shows that a change-over to a rod-like receiving area has little effect on the characteristics of the omni-directional wireless channel. An electrically small rod yields 6 spatial subchannels. With a rod whose length is three times the wavelength, the number of spatial subchannels increases monotonically with an augment of SNR. The diminished number of the subchannels and the decreased capacity in Fig. 7.9 in comparison to those in Fig. 7.7 for $r = \lambda_w$ is accounted for by the diminished size of the receiving area.

The results presented in Fig. 7.10 – 7.12 have been obtained for the same parameters as in Fig. 7.7 – 7.9, except that the azimuth channel was considered and not the omnidirectional one. It was assumed that 59 horizontally polarized wave and 59 vertically polarized wave sources were positioned in the horizontal plane ($\theta = \pi/2$).

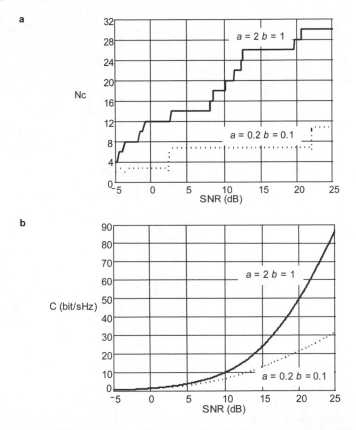

Fig. 7.10. SNR dependence of the 3-D azimuth channel characteristics for cylinder-shaped receiving areas. The cylinder height a and base radius b are expressed in wavelengths. The number of wave sources is 59 for each wave polarization type, the number of discrete points on the cylinder surface registering the electric and magnetic fields equals 375. Chart **a** represents the optimal number of spatial subchannels. Chart **b** represents the maximum capacity.

Figure 7.10 demonstrates the SNR dependence plots for the optimal number of subchannels and the channel limit capacity for a cylinder whose height a is equal to the base diameter ($a = 2b$). Under consideration are the electrically small cylinder ($a = 0.2\lambda_w$, $b = 0.1\lambda_w$) and the cylinder of the dimensions commensurate with the wavelength ($a = 2\lambda_w$, $b = \lambda_w$). Calculations were based on the assumption that the fields are recorded at 375 points over the cylinder surface.

Figure 7.11 illustrates the characteristics of the 3-D azimuth channel with a disk-shaped cylinder receiving area. Under examination are the small-height cylinders ($a = 0.1\lambda_w$, $b = \lambda_w$) and ($a = 0.01\lambda_w$, $b = 0.1\lambda_w$). The number of the field pick-up points is 500.

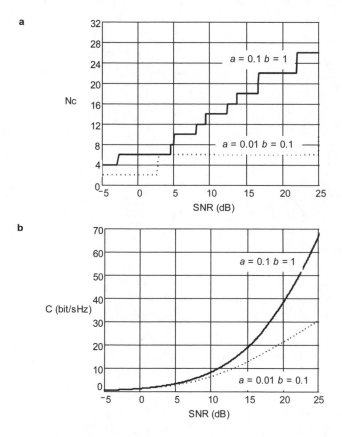

Fig. 7.11. SNR dependence of the characteristics of the 3-D azimuth channel with cylindrical receiving areas. The cylinder height a and base radius b are shown in wavelengths. The number of wave sources of each wave polarization type is 59, the number of electric and magnetic field pick-up discrete points on the surface of the cylinder is 500. Chart **a** represents the optimal number of spatial subchannels. Chart **b** represents the maximum capacity.

Figure 7.12 demonstrates the results of analyzing the 3-D azimuth channel with a rod-like cylinder receiving area. Under consideration are the cylinders ($a = 3\lambda_w$, $b = 0.01\lambda_w$) and ($a = 0.3\lambda_w$, $b = 0.001\lambda_w$) extended along the z-axis. The number of field pick-up points is taken to be equal 61.

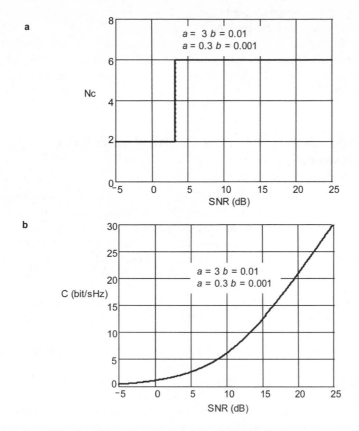

Fig. 7.12. SNR dependence of the characteristics of the 3-D azimuth channel for cylinder shaped receiving areas. The cylinder height a and base radius b are shown in wavelengths. The number of wave sources of each wave polarization type is 59, the number of electric and magnetic field recording discrete points on the surface of the cylinder is 61. Chart **a** represents the optimal number of spatial subchannels. Chart **b** represents the maximum capacity.

An examination of the plots in Fig. 7.10a, 7.11a and 7.12a reveals a close agreement of the plots for the small-size receiving areas. All the plots show that the optimal number of spatial subchannels is variable from 2 to 6 with SNR ≈ 3-DB. The availability of just two spatial subchannels and small SNR values makes the azimuth channel similar to the 2-D channel.

The plots of Fig. 7.10 – 7.12 disclose that in the azimuth channel with a stable SNR, a size increase in the vertical direction leads to practically zero rise in the optimal number of spatial subchannels and the limit capacity. An augmented size in the horizontal plane is found more effective. Figure 7.12 is most illustrative in this perspective. An extension of the a length of the rod-like receiving area from $0.3\lambda_w$ to $3\lambda_w$ has no effect on the channel characteristics; the plots representing different parameters of the receiving area almost coincide in Fig. 7.12.

Compare the solid curves in Fig. 7.10 and 7.11 representing various height and identical base radius values. The comparison shows that a decrease of the cylindrical area height by a factor of 20 (from $2\lambda_w$ to $0.1\lambda_w$) changes the channel characteristics insignificantly.

7.3 Three-dimensional Wireless Channel with a Box-Shaped Receiving Area

Let us turn now to the 3-D wireless channel with a parallelepiped receiving area. The virtue of this receiving area geometry is that its dimension in any of the three directions can be arbitrary. It is assumed that the parallelepiped faces are aligned parallel to the coordinate planes. For the purpose of the channel matrix estimation, it is assumed that the discrete points on the surface of the parallelepiped register the Cartesian components of the electric and magnetic fields. In other words, we make use of the relationships of Sect. 6.3.

The results of estimation are shown in Fig. 7.13–7.15. Figure 7.13 depicts a three-dimensional omnidirectional channel. The angular coordinates θ and φ of the wave sources are derived by formulae (6.10); the number of the wave sources of each wave polarization type was taken to be equal to 59. The parallelepiped dimensions are a, b and c along the x, y and z axes respectively.

Figure 7.13 demonstrates the characteristics of the 3-D omnidirectional channel. The dimensions of the parallelepiped vary so that its volume remains unchanged. The number of discrete points on the surface that record the electric and magnetic fields varies and equals 152, 236 and 260 for the sizes $a = b = c = \lambda_w$; $a = b = 2\lambda_w$, $c = 0.25\lambda_w$; $a = 4\lambda_w$, $b = \lambda_w$, $c = 0.25\lambda_w$ respectively.

A comparison of the curves in Fig. 7.13 reveals that in the 3-D omnidirectional channel the variation of its characteristics brought about by changes in the dimensions of the parallelepiped with its volume remaining constant is minor.

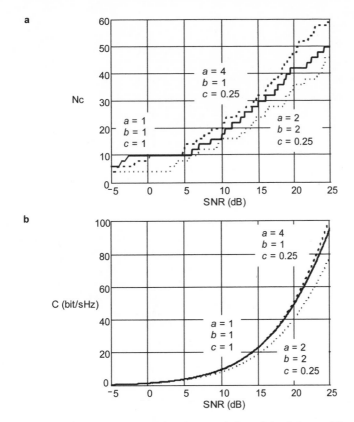

Fig. 7.13. SNR dependence of the characteristics of the 3-D omnidirectional channel for box-shaped receiving areas. The box dimensions a, b and c along the x, y and z axes are presented in wavelengths. The number of sources of each wave polarization type is 59, the number of electric and magnetic field pick-up discrete points on the surface of the parallelepiped varies and equals 152, 236 and 260 for the sizes $a = b = c = 1$; $a = b = 2, c = 0.25$; $a = 4, b = 1, c = 0.25$ respectively. Chart **a** represents the optimal number of spatial subchannels. Chart **b** represents the maximum capacity.

Figure 7.14 and 7.15 demonstrate the characteristics of the azimuth channel. Figure 7.14 illustrates a parallelepiped shaped as a bar. The bar positioned in the horizontal plane ($a = b = 2\lambda_w$, $c = 0.25\lambda_w$) and one positioned in the vertical plane ($a = c = 2\lambda_w$, $b = 0.25\lambda_w$) are considered. The number of the discrete points recording the electric and magnetic fields is 236.

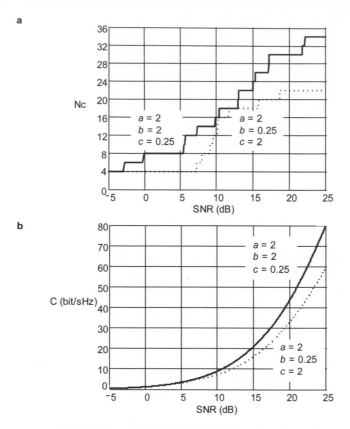

Fig. 7.14. SNR dependence of the characteristics of the 3-D azimuth channel for box-shaped receiving areas (bar). The parallelepiped dimensions a, b and c along the x-, y- and z-axes are given in wavelengths. The number of wave sources of each wave polarization type is 59, the number of electric and magnetic field pick-up discrete points on the surface of the parallelepiped is 236. Chart **a** represents the optimal number of spatial subchannels. Chart **b** represents the maximum capacity.

The plots in Fig. 7.14 indicate that in the azimuth channel, positioning the bar in the horizontal plane allows for a greater number of spatial subchannels and a greater limit capacity than its placing in the vertical plane.

Figure 7.15 deals with the azimuth wireless channel with the receiving area shaped as an elongated parallelepiped. The characteristics of the horizontally elongated parallelepiped ($a = 5\lambda_w$, $b = c = 0.25\lambda_w$) and vertically elongated one ($a = b = 0.25\lambda_w$, $c = 5\lambda_w$) are compared. The number of electric and magnetic field pick-up discrete points on the surface of the parallelepiped is 178.

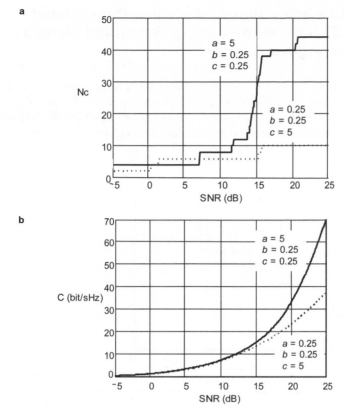

Fig. 7.15. SNR dependence of the characteristics of the 3-D azimuth channel for box-shaped receiving areas (rod-like parallelepiped). The parallelepiped dimensions a, b and c along the x-, y- and z-axes are given in wavelengths. The number of wave sources of each wave polarization type is 59, the number of electric and magnetic field pick-up discrete points on the surface of the parallelepiped is 178. Chart **a** represents the optimal number of spatial subchannels. Chart **b** represents the maximum capacity.

The plots of Fig. 7.15 make it apparent that in the azimuth wireless channel, an extension in the horizontal dimension is more advantageous for a gain in the number of spatial subchannels and the limit capacity than vertical elongation. It is worth noting that the advantages of one receiving area over the other as well as the merits of space coding communication systems become most pronounced with big SNR values.

7.4 Limiting Characteristics of the Three-Dimensional Wireless Channel with Potential Recording Along a Rectilinear Segment

Let us consider a three-dimensional wireless channel where reception means recording potentials along a rectilinear segment. The segment is held to be positioned along the z-axis. It is further assumed that the potentials of the electric and magnetic fields are recorded at discrete points along the segment. That is, in computations we make use of the formulae of Sect. 6.2.

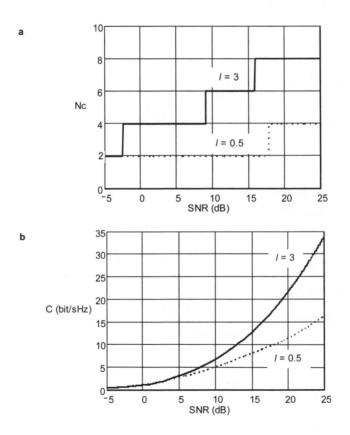

Fig. 7.16. SNR dependence of the characteristics of the 3-D omnidirectional channel for the receiving areas shaped as a vertical rod. The rod length l is given in wavelengths. The number of wave sources of each wave polarization type is 59, the number of discrete points recording the electric and magnetic potentials is 100. Chart **a** represents the optimal number of spatial subchannels. Chart **b** represents the maximum capacity.

Figure 7.16 shows the characteristics of the 3-D omnidirectional channel. The angular coordinates θ and φ of the wave sources are derived by formulae 6.10. The number of wave sources in the calculations was taken to be 59. The segments of the $l = 3\lambda_w$ and $l = 0.5\lambda_w$ have been considered. The number of discrete points along the segment is 100.

It is instructive to compare the characteristics presented in Fig. 7.16 for the vertical rod with those in Fig. 7.9 for the rod-like cylinder. The comparison discloses that the limit capacity for the rod-like cylinder is materially bigger than that for the rod. With the rod-like cylinder the optimal number of spatial subchannels is roughly 3 times as that with the rod. The following physical explanation can be supplied for the fact. Recording potentials along the rectilinear segment permits determination of the vertical components of the electric and magnetic fields only. While analyzing the cylinder-shaped receiving areas, it was assumed that all three components of the electric and magnetic fields are picked up. A greater number of recorded components gives a bigger number of independent spatial subchannels for the cylindrical receiving area and a bigger limit capacity.

Note the following fact that becomes apparent from examining the plot in Fig. 7.16a for $l = 0.5\lambda_w$. The linear area of the most commonly used size, $0.5\lambda_w$, with a not super-big SNR (below 18 dB), does not allow setting up communication systems with spatial subchannels – the optimal number of subchannels being equal to two even in the omnidirectional wireless channel. This means that only polarization diversity is possible. Creation of systems with spatial subchannels in excess of two requires either elongation of the linear area or transition to a 3-D receiving area.

Figure 7.17 presents the characteristics of the 3-D azimuth wireless channel. The number of wave sources was taken to be equal to 59 with all of them positioned in the horizontal plane. As in Fig. 7.16, the segments of the $l = 3\lambda_w$ and $l = 0.5\lambda_w$ length were considered. The number of discrete points along the segment is 100.

As is obvious from Fig. 7.17 the plots for the receiving areas of different sizes ($l = 3\lambda_w$ and $l = 0.5\lambda_w$) coincide. The number of spatial subchannels is not dependent on SNR or the rod length, and equals two. Thus, only polarization diversity is possible for the two subchannels of the azimuth wireless channel with a zero horizontal dimension of the receiving area.

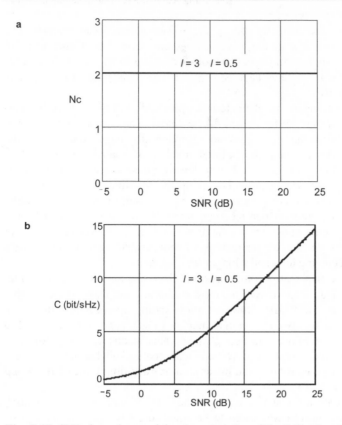

Fig. 7.17. SNR dependence of the characteristics of the 3-D azimuth wireless channel for the receiving areas shaped as a vertical rod. The l size of the rod is shown in wavelengths. The number of wave sources of each wave polarization type is 59, the number of discrete points at which electric and magnetic potentials are recorded is 100. Chart **a** represents the optimal number of spatial subchannels. Chart **b** represents the maximum capacity.

7.5 Summary

Chapter 7 presents the limit characteristics of the 3-D multipath wireless channel with a receiving area of various shapes.

1. For the 3-D omnidirectional wireless channel with an electrically small receiving area, it has been demonstrated that the optimal number of spatial subchannels is equal to 6. Therefore, to attain a channel capacity close to the limiting one with a moderate SNR value (SNR < 5dB), no spatial separation of the receive antenna elements is required. Fundamentally, the multipath 3-D channel permits employment of electrically small receiving antennas that secure construction of communication systems with six spatial subchannels.

2. In the three-dimensional omnidirectional wireless channel the optimal number of spatial subchannels rapidly increases with a growth of the 3-D receiving area size. This makes it possible to attain a capacity close to the limiting one $C_\infty =$ SNR/ln2 with moderate sizes of antennas (roughly $2\lambda_w$). In other words, the 3-D omnidirectional wireless channel permits development of high spectral efficiency antennas whose dimensions are comparable to the wavelength.
3. A limitation on space arrangement of the wave sources, a transition from the omnidirectional channel to the azimuth or sector one decreases the optimal number of spatial subchannels and the limit capacity of the wireless channel.
4. Investigation of a variety of receiving areas in the omnidirectional wireless channel reveals that it is not so much the shape of the receiving area that is important for building multichannel communication systems as it is the volume. The linear area of a zero volume receiving area allows for a less number of spatial subchannels than 3-D areas.
5. In the azimuth wireless channel an increase of the receiving area size in the vertical dimension causes little change, if any, in the optimal number of spatial subchannels, with a steady SNR. An increase of the receiving area dimension in the horizontal plane leads to a gain in the optimal number of spatial subchannels in the azimuth channel and the limit capacity of the communication system.

8 Statistical Model of the Three-Dimensional Multipath Wireless Channel

Three-dimensional wireless channel modeling has received less extensive coverage in literature than two-dimensional modeling. With a traditional ray-based approach to modeling [3] the resulting three-dimensional model appears much more cumbersome than a two-dimensional one. At the same time, a two-dimensional model lacks accuracy in terms of spatial treatment since it does not take into consideration vertical field variations.

The model of a three-dimensional wireless channel described in this chapter is built with the same methods that were used in modeling the two-dimensional wireless channel covered in Sect. 4.4. Electric and magnetic potentials in three-dimensional space are presented as a sum of basis functions that make spherical solutions to Maxwell's equations. All the components of the electromagnetic field are determined in terms of the coefficients $\alpha_{n,l}$ ($\beta_{n,l}$). The statistical properties of the random coefficients define the statistical properties of the model.

In Sect. 8.1 we determine the statistical characteristics of the model coefficients. The modeling results are given in Sect. 8.2.

8.1 Statistical Properties of the Model Coefficients

Spherical solutions to Maxwell's equations can provide convenient basis functions for three-dimensional modeling. Writing the potentials and fields in terms of basis functions is outlined in Chap. 5 and is represented by equations (5.2, 5.3, 5.5, and 5.7)

In order to determine the statistical properties of the coefficients included in these formulae, let us make use of their relationship (5.85-5.88) to the complex amplitudes of plane waves. In formulae (5.85-5.88) as well as in Chap. 6, the characteristic impedance of the medium $(\mu/\varepsilon)^{0.5}$ is taken as 1. The $(\mu/\varepsilon)^{0.5} = 1$ assumption means that detecting the magnetic field of a plane wave is identical to detecting the electric field of the wave. Furthermore, for the purpose of brevity, we will use the following designation:

$$f(n,l) = \frac{2n+1}{n(n+1)} \frac{(n-l)!}{(n+l)!} \qquad (8.1)$$

Then expressions (5.85-5.88) for the coefficients of spherical harmonics will be written as:

$$
\alpha_{n,l} = i^{n-1} f(n,l) \sum_{m=0}^{M-1} \frac{d\left(P_n^{(l)}(\cos\theta)\right)}{d\theta}\bigg|_{\theta=\theta_m} e^{-il\varphi_m} E_m^v -
$$

$$
- i^n lf(n,l) \sum_{m=0}^{M-1} \frac{P_n^{(l)}(\cos\theta_m)}{\sin\theta_m} e^{-il\varphi_m} E_m^h
$$

(8.2a)

$$
\beta_{n,l} = i^n lf(n,l) \sum_{m=0}^{M-1} \frac{P_n^{(l)}(\cos\theta_m)}{\sin\theta_m} e^{-il\varphi_m} E_m^v +
$$

$$
+ i^{n-1} f(n,l) \sum_{m=0}^{M-1} \frac{d\left(P_n^{(l)}(\cos\theta)\right)}{d\theta}\bigg|_{\theta=\theta_m} e^{-il\varphi_m} E_m^h
$$

(8.2b)

In formulae (8.2) the complex amplitudes of the vertically and horizontally polarized waves are represented by E_m^v, E_m^h respectively. The θ_m, φ_m angles in spherical coordinates determine the direction to the $m\text{-}th$ wave source.

It is assumed that the complex amplitudes of wave sources are independent stochastic variables with zero mean, therefore their statistical properties may be defined by the following relationships:

$$
\left\langle E_m^v \right\rangle = \left\langle E_m^h \right\rangle = 0
$$

(8.3a)

$$
\left\langle E_{m2}^v E_{m1}^{v*} \right\rangle = \left\langle E_{m2}^h E_{m1}^{h*} \right\rangle = 0, \text{ with } m2 \neq m1
$$

(8.3b)

$$
\left\langle \left| E_m^v \right|^2 \right\rangle = D_m^v, \ \left\langle \left| E_m^h \right|^2 \right\rangle = D_m^h
$$

(8.3c)

$$
\left\langle E_{m2}^v E_{m1}^{h*} \right\rangle = 0
$$

(8.3-D)

The number of wave sources M is deemed big enough to permit introduction of the continuous functions $d^v(\theta,\varphi)$ and $d^h(\theta,\varphi)$, representing the angular distribution of power for vertically and horizontally polarized waves. The power of the waves from the sources confined to small angles $\Delta\theta$, $\Delta\varphi$ is

$$
d^v(\theta,\varphi)\sin\theta\Delta\theta\Delta\varphi
$$

(8.4a)

$$
d^h(\theta,\varphi)\sin\theta\Delta\theta\Delta\varphi
$$

(8.4b)

The overall power of all the sources of a single polarization can be calculated by:

$$\frac{1}{2}\sum_{m=0}^{M-1} D_m^v = \int_0^{2\pi}\int_0^{\pi} d^v(\theta,\varphi)\sin\theta d\theta d\varphi \tag{8.5a}$$

$$\frac{1}{2}\sum_{m=0}^{M-1} D_m^h = \int_0^{2\pi}\int_0^{\pi} d^h(\theta,\varphi)\sin\theta d\theta d\varphi \tag{8.5b}$$

As in the case of the two-dimensional wireless channel we will look now at three cases of angular arrangement of the wave sources. Case 1 is a homogeneous distribution of the sources in all the dimensions of a three-dimensional space. Case 2 presents an inhomogeneous distribution of random sources. Case 3 includes a nonuniform distribution of random sources in combination with a deterministic plane wave.

Case 1. Homogeneous distribution of wave sources on a spherical surface
We assume, that in Case 1 the variance of complex amplitudes in all waves is identical, that is (8.3c) may be written as

$$D_m^v = D_m^h = \left\langle |E_0|^2 \right\rangle \tag{8.6}$$

The functions $d^v(\theta,\varphi)$ and $d^h(\theta,\varphi)$ of Case 1 must be deemed constant. From (8.5) follows their relation to the variance $<|E_0|^2>$ of a single source.

$$d^v(\theta,\varphi) = d^h(\theta,\varphi) = d = \frac{M}{8\pi}\left\langle |E_0|^2 \right\rangle \tag{8.7}$$

Let us determine the statistical properties of the complex amplitudes $\alpha_{n,l}$ ($\beta_{n,l}$) of spherical harmonics.

Averaging (8.2) in view of (8.3a), we ascertain that the mean values of the complex amplitudes are zero.

$$\left\langle \alpha_{n,l} \right\rangle = \left\langle \beta_{n,l} \right\rangle = 0 \tag{8.8}$$

We will now calculate covariance for various coefficients. The computations based on (8.2a) in terms of (8.3b), (8.3-D) and (8.6) yield the following covariance expression.

$$\left\langle \alpha_{n2,l2}\alpha_{n1,l1}^* \right\rangle = i^{n2-n1} f(n_1,l_1)f(n_2,l_2)\left\langle |E_0|^2 \right\rangle \times$$

$$\times \sum_{m=0}^{M-1}\left(\frac{d\left(P_{n1}^{(l1)}(\cos\theta)\right)}{d\theta}\bigg|_{\theta=\theta_m} \frac{d\left(P_{n2}^{(l2)}(\cos\theta)\right)}{d\theta}\bigg|_{\theta=\theta_m} + l_1 l_2 \frac{P_{n1}^{(l1)}(\cos\theta_m)P_{n2}^{(l2)}(\cos\theta_m)}{\sin^2\theta_m} \right) e^{i(l1-l2)\varphi_m} \tag{8.9}$$

Assuming that the number of wave sources is big we will substitute an integral for the summation in (8.9). In doing so the angular distribution density d should be used in lieu of the single source variance $<|E_0|^2>$.

$$\left\langle \alpha_{n2,l2}\alpha^*_{n1,l1}\right\rangle = 2i^{n2-n1} f(n_1,l_1)f(n_2,l_2)d\int_0^{2\pi} e^{i(l1-l2)\varphi}\,d\varphi \times$$

$$\times \int_0^{\pi}\left(\sin\theta \frac{d\left(P_{n1}^{(l1)}(\cos\theta)\right)}{d\theta}\frac{d\left(P_{n2}^{(l2)}(\cos\theta)\right)}{d\theta} + l_1 l_2 \frac{P_{n1}^{(l1)}(\cos\theta)P_{n2}^{(l2)}(\cos\theta)}{\sin\theta}\right)d\theta \qquad (8.10)$$

The integral in (8.10) was evaluated during power estimation in Sect. 5.4. It is zero due to orthogonality of the spherical harmonics when either of the indices are not equal ($n1 \neq n2$ or $l1 \neq l2$). The integral will be determined by (5.30) if the indices are equal. Use of (5.30) will allow us to write:

$$\left\langle \alpha_{n2,l2}\alpha^*_{n1,l1}\right\rangle = 0 \text{, if } n1 \neq n2 \text{ or } l1 \neq l2 , \qquad (8.11a)$$

$$\left\langle \left|\alpha_{n,l}\right|^2\right\rangle = 8\pi d f(n,l)$$

Absolutely identical rearrangements result in exactly the same expressions for the coefficients $\beta_{n,l}$:

$$\left\langle \beta_{n2,l2}\beta^*_{n1,l1}\right\rangle = 0 \text{, if } n1 \neq n2 \text{ or } l1 \neq l2 , \qquad (8.11b)$$

$$\left\langle \left|\beta_{n,l}\right|^2\right\rangle = 8\pi d f(n,l)$$

Let us consider the covariance of the coefficients α and β. Equations (8.2) lead to the following expression:

$$\left\langle \alpha_{n2,l2}\beta^*_{n1,l1}\right\rangle = i^{n2-n1-1} f(n_1,l_1)f(n_2,l_2)\left\langle \left|E_0\right|^2\right\rangle \times$$

$$\times \sum_{m=0}^{M-1}\left(l_2 \frac{d\left(P_{n1}^{(l1)}(\cos\theta)\right)}{d\theta}\bigg|_{\theta=\theta_m} \frac{P_{n2}^{(l2)}(\cos\theta_m)}{\sin\theta_m} + l_1 \frac{P_{n1}^{(l1)}(\cos\theta_m)}{\sin\theta_m} \frac{d\left(P_{n2}^{(l2)}(\cos\theta)\right)}{d\theta}\bigg|_{\theta=\theta_m}\right)e^{i(l1-l2)\varphi_m}$$

In substitution of integral for summation we take into account that the integral of $\exp(i(l_1-l_2)\varphi)$ taken over φ is nonzero only when $l_1 = l_2 = l$. Hence we obtain:

$$\left\langle \alpha_{n2,l}\beta^*_{n1,l}\right\rangle = 4\pi i^{n2-n1-1} lf(n_1,l)f(n_2,l)d \times$$

$$\times \int_0^{\pi}\left(\frac{d\left(P_{n1}^{(l)}(\cos\theta)\right)}{d\theta}P_{n2}^{(l)}(\cos\theta) + P_{n1}^{(l)}(\cos\theta)\frac{d\left(P_{n2}^{(l)}(\cos\theta)\right)}{d\theta}\right)d\theta$$

An integration by parts yields the following covariance expression:

$$\left\langle \alpha_{n2,l}\beta^*_{n1,l}\right\rangle = 4\pi i^{n2-n1-1} lf(n_1,l)f(n_2,l)d \left. P_{n1}^{(l)}(\cos\theta)P_{n2}^{(l)}(\cos\theta)\right|_{\theta=0}^{\theta=\pi} \qquad (8.12)$$

The formula for the associated Legendre functions [19] given below

$$P_n^{(l)}(\cos\theta) = \left(\cos\frac{\theta}{2}\right)^l \sum_{j=l}^{n} \frac{(-1)^j (n+j)!}{(n-j)!(j-l)!\,j!}\left(\sin\frac{\theta}{2}\right)^{2j-l}$$

shows that when $l \geq 1$, the associated Legendre function contains multiplier $\sin\theta$. Therefore, whatever the indices are, from (8.12) it follows

$$\left\langle \alpha_{n2,l2}\beta_{n1,l1}^{*}\right\rangle = 0 \tag{8.13}$$

It is apparent then that the complex amplitudes $\alpha_{n,l}$ and $\beta_{n,l}$ of spherical harmonics in the three-dimensional omnidirectional wireless channel are statistically independent normal random quantities with zero mean values. Their variance can be calculated by formulae (8.11). The normal distribution law follows from the central limit theorem of the probability theory.

While deriving (8.11) it was assumed that writing the fields in terms of formulae (5.5) and (5.7) does not involve normalization of the associated Legendre functions. In what follows, we will use normalized associated Legendre functions to reduce (8.11). While normalizing, the multiplier (6.1) is introduced into the Legendre function. Owing to this, the variance expression (8.11) should be divided by the squared multiplier (6.1). In view of designation (8.1) we obtain the following variance expression for the coefficients $\alpha N_{n,l}$ and $\beta N_{n,l}$ during normalization

$$\left\langle \left|\alpha N_{n,l}\right|^2\right\rangle = \left\langle \left|\beta N_{n,l}\right|^2\right\rangle = 16\pi^2 d \tag{8.14}$$

This means that in the omnidirectional wireless channel, with normalization of the associated Legendre functions, the complex amplitudes of all spherical harmonics have the same variance. It differs from the angular power density d by the multiplicative constant $16\pi^2$. Relationship (8.7) permits us to express the variance of the coefficients in terms of the total number of sources and the complex amplitude variance of a single source.

$$\left\langle \left|\alpha N_{n,l}\right|^2\right\rangle = \left\langle \left|\beta N_{n,l}\right|^2\right\rangle = 2\pi M\left\langle \left|E_0\right|^2\right\rangle \tag{8.15}$$

Even more appropriate than (8.14) and (8.15) would be expressing the coefficient variance in terms of the variance of one of the electric field components. Let us calculate the variance for one (say, the vertical) component of the field

$$\left\langle \left|E_z\right|^2\right\rangle = \left\langle \left|E_0\right|^2\right\rangle \sum_{m=0}^{M-1} \sin^2\theta_m$$

Substituting integral for the summation we obtain

$$\left\langle \left|E_z\right|^2\right\rangle = 2d \int_0^{2\pi}\int_0^{\pi} \sin^3\theta d\theta$$

Evaluating the integral we have

$$\left\langle \left|E_z\right|^2\right\rangle = \frac{16}{3}\pi d \tag{8.16}$$

In terms of (8.16) expression (8.14) assumes the form

$$\left\langle \left|\alpha N_{n,l}\right|^2\right\rangle = \left\langle \left|\beta N_{n,l}\right|^2\right\rangle = 3\pi\left\langle \left|E_z\right|^2\right\rangle \tag{8.17}$$

Expression (8.17) determines the variance of all the coefficients $\alpha N_{n,l}$ and $\beta N_{n,l}$ in terms of the variance of the vertical component of the electric field. With unit variance (8.17) yields

$$\left\langle \left|\alpha N_{n,l}\right|^2\right\rangle = \left\langle \left|\beta N_{n,l}\right|^2\right\rangle = 3\pi \tag{8.17a}$$

Let us now summarize the review of the three-dimensional omnidirectional wireless channel and outline its modeling procedure.

1. Generation of the coefficients $\alpha N_{n,l}$ and $\beta N_{n,l}$

In case of an omnidirectional wireless channel, all the coefficients $\alpha N_{n,l}$ and $\beta N_{n,l}$ are independent normal random quantities with zero mean and equal variances. The variance of each coefficient is related to the variance of one of the components of the electric field through (8.17). In case of the component unit variance, the variance of each coefficient is 3π (8.17a).

2. Calculation of fields for various spatial points

The fields are calculated by formulae (5.5 and 5.7). We will now present the estimation relationships in view of normalization of the associated Legendre functions and with the understanding that the characteristic impedance of the medium is unity.

$$E_r(r,\theta,\varphi) = \sum_{n=1}^{N\max}\sum_{l=-n}^{n}\alpha N_{n,l}\left(\frac{d^2\left(krj_n(kr)\right)}{d(kr)^2} + krj_n(kr)\right)PN_n^{(l)}(\cos\theta)e^{il\varphi} \tag{8.18a}$$

$$E_\theta(r,\theta,\varphi) = \frac{1}{kr}\sum_{n=1}^{N\max}\sum_{l=-n}^{n}\alpha N_{n,l}\frac{d\left(krj_n(kr)\right)}{d(kr)}\frac{d\left(PN_n^{(l)}(\cos\theta)\right)}{d\theta}e^{il\varphi} + \tag{8.18b}$$

$$+\frac{1}{\sin\theta}\sum_{n=1}^{N\max}\sum_{l=-n}^{n}\beta N_{n,l}\, lj_n(kr)PN_n^{(l)}(\cos\theta)e^{il\varphi}$$

$$E_\varphi(r,\theta,\varphi) = \frac{i}{kr\sin\theta}\sum_{n=1}^{N\max}\sum_{l=-n}^{n}\alpha N_{n,l}\,l\frac{d\left(krj_n(kr)\right)}{d(kr)}PN_n^{(l)}(\cos\theta)e^{il\varphi} + \tag{8.18c}$$

$$+i\sum_{n=1}^{N\max}\sum_{l=-n}^{n}\beta N_{n,l}\,j_n(kr)\frac{d\left(PN_n^{(l)}(\cos\theta)\right)}{d\theta}e^{il\varphi}$$

$$H_r(r,\theta,\varphi) = \sum_{n=1}^{N\max}\sum_{l=-n}^{n}\beta N_{n,l}\left(\frac{d^2\left(krj_n(kr)\right)}{d(kr)^2} + krj_n(kr)\right)PN_n^{(l)}(\cos\theta)e^{il\varphi} \tag{8.19a}$$

$$H_\theta(r,\theta,\varphi) = -\frac{1}{\sin\theta} \sum_{n=1}^{N\max} \sum_{l=-n}^{n} \alpha N_{n,l} l j_n(kr) PN_n^{(l)}(\cos\theta) e^{il\varphi} + \qquad (8.19b)$$

$$+\frac{1}{kr} \sum_{n=1}^{N\max} \sum_{l=-n}^{n} \beta N_{n,l} \frac{d\big(krj_n(kr)\big)}{d(kr)} \frac{d\big(PN_n^{(l)}(\cos\theta)\big)}{d\theta} e^{il\varphi}$$

$$H_\varphi(r,\theta,\varphi) = -i \sum_{n=1}^{N\max} \sum_{l=-n}^{n} \alpha N_{n,l} j_n(kr) \frac{d\big(PN_n^{(l)}(\cos\theta)\big)}{d\theta} e^{il\varphi} + \qquad (8.19c)$$

$$+\frac{i}{kr\sin\theta} \sum_{n=1}^{N\max} \sum_{l=-n}^{n} \beta N_{n,l} l \frac{d\big(krj_n(kr)\big)}{d(kr)} PN_n^{(l)}(\cos\theta) e^{il\varphi}$$

In formulae (8.18-8.19) r, θ and φ are the spherical coordinates of the observation point. $k = 2\pi/\lambda_w$ is the wave number. N_{max} denotes the maximal value of the first index of a spherical harmonic. The value selected for N_{max} should be about one order of magnitude greater than r_{max}/λ_w. r_{max} signifies the radius of the area within the confines of which formulae (8.18-8.19) hold true. $j_n(kr)$ stands for the spherical Bessel function, determined in Chap. 5 and calculated by formulae (5.9) or (5.12). $PN_n^{(l)}(\cos\theta)$ is the normalized associated Legendre function. It is related to the non-normalized function $P_n^{(l)}(\cos\theta)$ by the relationship

$$PN_n^{(l)}(\cos\theta) = \sqrt{\frac{1}{2\pi} \frac{2n+1}{n(n+1)} \frac{(n-l)!}{(n+l)!}} P_n^{(l)}(\cos\theta) \qquad (8.20)$$

The associated Legendre function may be calculated by (5.14). All the electric and magnetic field components are defined by relationships (8.18 - 8.19).

3. Calculation of the signal at the antenna output

Items 1) and 2) present a model of a 3-D electromagnetic field. This model may be complemented by a model of an antenna system, which would allow determination of the signals at its output from the known field. With the modeling results obtained as shown in Sect. 8.2, it is assumed that the antenna elements are field sensors. To put it differently, we assume that the antenna system detects various Cartesian field components at a variety of spatial points.

As far as the Doppler shift is concerned it makes sense to repeat exactly what was stated for the 2-D wireless channel model, namely that the Doppler shift will automatically appear during calculation of the signal at the output of the receiving antenna in motion. A space-variant field will cause a time-variant signal at the output of the moving antenna. The spectrum of this signal will include the Doppler frequency shift.

Case 2. Inhomogeneous distribution of wave sources

It is assumed that as distinguished from Case 1 the statistical characteristics of the random sources in Case 2 are determined by general expressions (8.3). In exactly the same way as in Case 1 it is easy to establish that the complex amplitude mean of each spherical harmonic is zero

$$\left\langle \alpha_{n,l} \right\rangle = \left\langle \beta_{n,l} \right\rangle = 0 \tag{8.21}$$

Based on (8.2) and (8.3) we can write covariances for the complex amplitudes of spherical harmonics

$$\left\langle \alpha_{n2,l2}\alpha_{n1,l1}^{*} \right\rangle = i^{n2-n1} f(n_1,l_1)f(n_2,l_2) \times$$

$$\times \sum_{m=0}^{M-1} \left(\frac{d\left(P_{n1}^{(l1)}(\cos\theta)\right)}{d\theta}\bigg|_{\theta=\theta_m} \frac{d\left(P_{n2}^{(l2)}(\cos\theta)\right)}{d\theta}\bigg|_{\theta=\theta_m} \left\langle \left|E_m^v\right|^2 \right\rangle + \right. \left. + l_1 l_2 \frac{P_{n1}^{(l1)}(\cos\theta_m)P_{n2}^{(l2)}(\cos\theta_m)}{\sin^2\theta_m} \left\langle \left|E_m^h\right|^2 \right\rangle \right) e^{i(l1-l2)\varphi_m} \tag{8.22a}$$

$$\left\langle \beta_{n2,l2}\beta_{n1,l1}^{*} \right\rangle = i^{n2-n1} f(n_1,l_1)f(n_2,l_2) \times$$

$$\times \sum_{m=0}^{M-1} \left(l_1 l_2 \frac{P_{n1}^{(l1)}(\cos\theta_m)P_{n2}^{(l2)}(\cos\theta_m)}{\sin^2\theta_m} \left\langle \left|E_m^v\right|^2 \right\rangle + \right. \left. + \frac{d\left(P_{n1}^{(l1)}(\cos\theta)\right)}{d\theta}\bigg|_{\theta=\theta_m} \frac{d\left(P_{n2}^{(l2)}(\cos\theta)\right)}{d\theta}\bigg|_{\theta=\theta_m} \left\langle \left|E_m^h\right|^2 \right\rangle \right) e^{i(l1-l2)\varphi_m} \tag{8.22b}$$

$$\left\langle \alpha_{n2,l2}\beta_{n1,l1}^{*} \right\rangle = i^{n2-n1-1} f(n_1,l_1)f(n_2,l_2) \times$$

$$\times \sum_{m=0}^{M-1} \left(l_2 \frac{d\left(P_{n1}^{(l1)}(\cos\theta)\right)}{d\theta}\bigg|_{\theta=\theta_m} \frac{P_{n2}^{(l2)}(\cos\theta_m)}{\sin\theta_m} \left\langle \left|E_m^v\right|^2 \right\rangle + \right. \left. + l_1 \frac{P_{n1}^{(l1)}(\cos\theta_m)}{\sin\theta_m} \frac{d\left(P_{n2}^{(l2)}(\cos\theta)\right)}{d\theta}\bigg|_{\theta=\theta_m} \left\langle \left|E_m^h\right|^2 \right\rangle \right) e^{i(l1-l2)\varphi_m} \tag{8.22c}$$

In a similar manner the expressions can be written with integrals in lieu of summations

$$\left\langle \alpha_{n2,l2}\alpha_{n1,l1}^{*} \right\rangle = 2i^{n2-n1} f(n_1,l_1)f(n_2,l_2) \times$$

$$\times \int_0^{2\pi}\int_0^{\pi} \left(\sin\theta \frac{d\left(P_{n1}^{(l1)}(\cos\theta)\right)}{d\theta} \frac{d\left(P_{n2}^{(l2)}(\cos\theta)\right)}{d\theta} d^v(\theta,\varphi) + \right. \left. + l_1 l_2 \frac{P_{n1}^{(l1)}(\cos\theta)P_{n2}^{(l2)}(\cos\theta)}{\sin\theta} d^h(\theta,\varphi) \right) e^{i(l1-l2)\varphi} d\theta d\varphi \tag{8.23a}$$

$$\left\langle \beta_{n2,l2}\beta_{n1,l1}^{*} \right\rangle = 2i^{n2-n1}f(n_1,l_1)f(n_2,l_2)\times$$

$$\times \int_{0}^{2\pi}\int_{0}^{\pi} l_1 l_2 \left(\frac{P_{n1}^{(l1)}(\cos\theta)P_{n2}^{(l2)}(\cos\theta)}{\sin\theta}d^{v}(\theta,\varphi) + \sin\theta\frac{d\left(P_{n1}^{(l1)}(\cos\theta)\right)}{d\theta}\frac{d\left(P_{n2}^{(l2)}(\cos\theta)\right)}{d\theta}d^{h}(\theta,\varphi) \right) e^{i(l1-l2)\varphi}d\theta d\varphi \tag{8.23b}$$

$$\left\langle \alpha_{n2,l2}\beta_{n1,l1}^{*} \right\rangle = 2i^{n2-n1-1}f(n_1,l_1)f(n_2,l_2)\times$$

$$\times \int_{0}^{2\pi}\int_{0}^{\pi} l_2 \left(\frac{d\left(P_{n1}^{(l1)}(\cos\theta)\right)}{d\theta}P_{n2}^{(l2)}(\cos\theta)d^{v}(\theta,\varphi) + l_1 P_{n1}^{(l1)}(\cos\theta)\frac{d\left(P_{n2}^{(l2)}(\cos\theta)\right)}{d\theta}d^{h}(\theta,\varphi) \right) e^{i(l1-l2)\varphi}d\theta d\varphi \tag{8.23c}$$

With arbitrary angular distribution of the random sources there are no further ways to reduce expressions (8.22, 8.23).

Therefore, in Case 2 the complex amplitudes of various spherical harmonics turn out to be correlated. Their covariances are determined by expressions (8.22) or (8.23). After deriving the coefficients $\alpha_{n,l}$ $\beta_{n,l}$ for field calculation, as in Case 1, equations (8.18-8.19) should be used.

Case 3. Inhomogeneous distribution of wave sources with presence of a deterministic wave.

Case 3 differs from Case 2 (or from Case 1) by presence of a deterministic plane wave. The presence of the deterministic wave determines nonzero mean values of the coefficients $\alpha_{n,l}$ $(\beta_{n,l})$. The mean values of these coefficients can be calculated by the complex amplitudes E^{v}_{f}, E^{h}_{f} of the deterministic wave with the help of expressions (8.2):

$$\left\langle \alpha_{n,l} \right\rangle = i^{n-1}f(n,l)\frac{d\left(P_{n}^{(l)}(\cos\theta)\right)}{d\theta}\bigg|_{\theta=\theta_f} e^{-il\varphi_f}E_{f}^{v} -$$

$$- i^{n}lf(n,l)\frac{P_{n}^{(l)}(\cos\theta_f)}{\sin\theta_f}e^{-il\varphi_f}E_{f}^{h} \tag{8.24a}$$

$$\left\langle \beta_{n,l} \right\rangle = i^{n}lf(n,l)\frac{P_{n}^{(l)}(\cos\theta_f)}{\sin\theta_f}e^{-il\varphi_f}E_{f}^{v} +$$

$$+ i^{n-1}f(n,l)\frac{d\left(P_{n}^{(l)}(\cos\theta)\right)}{d\theta}\bigg|_{\theta=\theta_f}e^{-il\varphi_f}E_{f}^{h} \tag{8.24b}$$

The characteristics of the random components in Case 3 are determined exactly as in Case 2 or Case 1.

8.2 Results of the Three-Dimensional Multipath Wireless Channel Modeling

The model of a three-dimensional multipath wireless channel described in Sect. 8.1 discloses the typical relationship existing between the electric and magnetic fields and the spatial coordinates. The plots of this section show the x-coordinate dependence of various Cartesian components of the electric field.

Figure 8.1 illustrates the x-coordinate dependence of the intensity of the electric field vertical component E_z. For the purposes of calculation, the number of spherical harmonics of each type was assumed to be 728. In other words, 728 complex random coefficients α and the same number of coefficients β were generated during modeling. The x-axis step was taken to be equal to $0.05\lambda_w$.

The two curves shown in Fig. 8.1 are plotted for the points spaced $0.1\lambda_w$ apart along the z axis. One additional comment is necessary. In the three-dimensional wireless channel, in contrast to the 2-D channel, variation of the z coordinate results in change of the field value. In Fig. 8.1 with small z-axis steps ($0.1\lambda_w$) the variations of E_z are minor. Strong correlation between the dotted and the solid curves is evident.

Figure 8.2 shows similar dependences plotted for a greater separation along the z-axis ($2\lambda_w$). No correlation of the solid and the dotted curves is noticeable here.

Figure 8.3 and 8.4 depict the x-axis dependence of the intensity of the electric field horizontal component E_y . As before, 728 spherical harmonics of every type were taken into account. The solid and the dotted curves correspond to different values of the z coordinate. In Fig. 8.3 the spacing is $0.1\lambda_w$, while in Fig. 8.4 it is $2\lambda_w$. The curves in Fig. 8.3 and 8.4 are similar to the curves in Fig. 8.1 and 8.2 respectively. With small spacing strong correlation between the solid and the dotted curves is apparent in Fig. 8.3. With the greater separation in Fig. 8.4 correlation becomes inconspicuous.

The intensity variations of all the three Cartesian components of the electric field (E_x , E_y ,E_z) following the change of the x coordinate are demonstrated in Fig. 8.5.

As previously, 728 spherical E-harmonics and the same number of H-harmonics were used in modeling.

No correlation between various components of the omnidirectional wireless channel is obvious in Fig. 8.5.

In conclusion it may be said that modeling the omnidirectional wireless channel in terms of spherical harmonics requires no additional field normalization. With condition (8.17a) met, the mean value of the E_z component intensity as well as of all the rest is 1 (0 dB).

The number of harmonics taken into account determines the dimensions of the space area for which the modeling results are true. With the area of radius $3\lambda_w$ it suffices to take into account 728 spherical harmonics of each type. An area of a smaller size requires a less number of harmonics, while with expanded area dimensions it is necessary to increase the number of spherical harmonics taken into consideration.

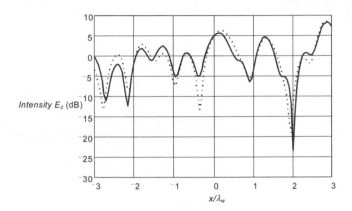

Fig. 8.1. Typical x-coordinate dependences of the intensity of the electric field vertical component. 728 spherical E- and H-harmonics were taken into account. Spacing along the z axis is $0.1\lambda_w$. The x-axis step is 0.05λ

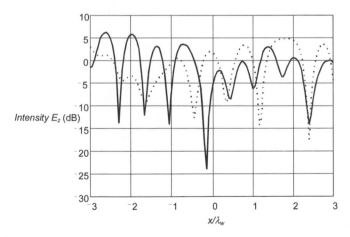

Fig. 8.2. Typical x-coordinate dependences of the intensity of the electric field vertical component. 728 spherical E- and H-harmonics were taken into account. Spacing along the z axis is $2\lambda_w$. The x-axis step $0.05\lambda_w$

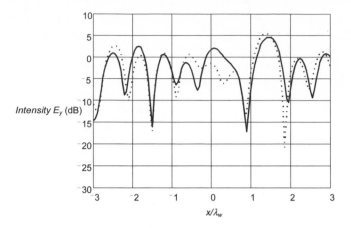

Fig. 8.3. Typical x-coordinate dependences of the intensity of the electric field horizontal component. 728 spherical E- and H-harmonics were taken into account. Spacing along the z axis is $0.1\lambda_w$. The x-axis step is $0.05\lambda_w$

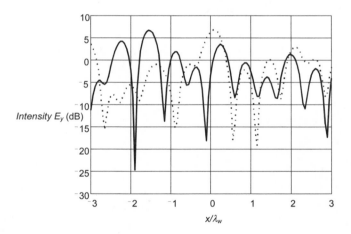

Fig. 8.4. Typical x-coordinate dependences of the intensity of the electric field horizontal component. 728 spherical E- and H-harmonics were taken into account. Spacing along the z axis is $2\lambda_w$. The x-axis step $0.05\lambda_w$

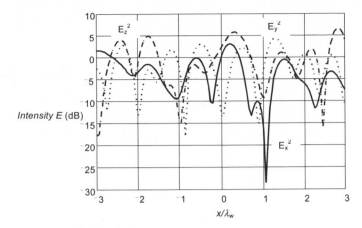

Fig. 8.5. Simultaneous measurement of the x-dependent intensities E_x, E_y and E_z of the three Cartesian components of an electric filed. 728 spherical E- and H-harmonics were taken into account. There is no spatial separation

8.3 Summary

A non-ray statistical model of the three-dimensional multipath wireless channel was described in Chap. 8. It presents an electromagnetic field in a limited area of a three-dimensional space as a set of spherical solutions to Maxwell's equations.

1. The statistical properties of the model coefficients have been determined. The model coefficients in case of an omnidirectional three-dimensional wireless channel have been demonstrated to be statistically independent normal random quantities with zero mean and identical variance. Such unsophisticated statistical properties indirectly confirm the adequacy of describing three-dimensional random fields in terms of spherical solutions to Maxwell's equations.

2. Analytical formulae determining the correlation properties of the model coefficients have been derived for a wireless channel that is not omnidirectional. Modeling such wireless channels requires prior generation of random coefficients with certain correlation properties.

3. Examples of omnidirectional channel modeling have been given. They provide support for the view that generation of various components of random electromagnetic fields for a variety of points in a three-dimensional space is possible.

Conclusion

In closing it may be said about the implications of the obtained results for development of communication systems. The graphs and computation expressions presented in the book define the limit data transfer rate that can be achieved in a multipath channel. The book presents the calculation techniques for determining the optimal number of spatial subchannels in a SDMA (spatial division multiple access) system, and the technology which secures this limiting data rate. The optimal number of spatial subchannels can remain large even if the dimensions of the antenna system are quite limited. The final gain in the data transfer rate as compared to that of a system without space diversity of its subchannels may prove to be enormous.

The results of Chap. 7, for example, demonstrate that in the 3-D omnidirectional channel with a spherical receiving area of the wavelength radius, for SNR = 20dB the limit capacity per unit bandwidth may approach 70 bit/s·Hz with the optimal number of spatial subchannels being 60.

Let us consider the abovementioned figures. According to Shannon the limit capacity of a SISO system with a 20dB SNR is 6.7 bit/s·Hz. In other words, the SDMA system is ten times as efficient as the SISO one, and is capable either of a tenfold data transfer rate or of a tenfold bandwidth economy with the same data transfer rate.

The 20dB SNR value taken into account in estimating the tenfold gain in performance may seem oversized and difficult to achieve. It should be remembered, though, that in this book the SNR is viewed as the signal to noise ratio with all the input power going into one (best performing) and the most efficient spatial subchannel. In a SDMA system the SNR value in each individual spatial subchannel will be less due to power distribution. Therefore, in the presented example, the SNR in each subchannel does not exceed 5 dB. Such an SNR value does not strike as oversized and is readily attainable.

Finally, the 60 spatial subchannels of the example discussed here are well worth another look. At present the SDMA communication systems are built around conventional antenna elements with spatial separation. In doing so it is believed that the separation should not be less than half the wavelength. This generally accepted concept is inconsistent with the presented figures. The best possible performance antenna must have 60 outputs, though the maximum spacing between the antenna elements must not exceed two wavelengths. This means that special antenna systems need to be developed for the limited-size SDMA communication systems. Conceivably these might be multielement antennas with peculiarly intertwined elements. An antenna system of this kind is likely to be a resonant system with a multitude of oscillation types, i.e., a multimode antenna.

The need for special types of antennas for SDMA communication systems is one of the essential conclusions of the book. The theoretical material presented in the book does not permit predictions as to the specific design of such antennas, but rather identifies an avenue of prospective research. Hopefully, the obtained results will attract the attention of antenna engineering experts whose expertise and skills will allow creation of innovative antenna systems for portable MIMO systems. Such antennas will no doubt enhance the efficiency of newly developed communication systems, bringing their performance characteristics close to the estimated attainable limits.

The book presents a new non-ray model of the random multipath channel. The statistical model of the 3-D wireless channel is based on the spherical harmonics. The electromagnetic field in a limited region of space is represented as a sum of the fields of spherical harmonics with random coefficients. For an omnidirectional wireless channel all the coefficients of the model are normal independent random quantities with zero mean values and identical variances. Such statistical properties of the coefficients suggest that the proposed model gives the most economical description of the coordinate dependence of the 3-D electromagnetic field.

Additionally, a non-ray model of the 2-D wireless channel has been proposed, based on the two-dimensional countertypes of the spherical harmonics.

The number of random coefficients required for the suggested model does not grow with an increase in the amount of multipath components and rapidly diminishes as the dimensions of the receiving area contract. The properties of the model put forward furnish a means for building simulations which can evaluate performance of portable communication systems designed for high levels of multipathing.

References

1. Jakes WC (1974) Microwave mobile communication. Wiley, New York

2. Lee WYC (1989) Mobile cellular telecommunications systems. McGraw Hill, New York

3. Parsons LD (1992) The mobile radio propagation channels. Pentech, New York

4. Rappaport TS (1996) Wireless communications. Prentice Hall, New York

5. Blaunstein N (1999) Radio propagation in cellular networks. Artech House, Boston

6. Liberti JC, Rappaport TS (1999) Smart antennas for wireless communications. Prentice Hall, New York

7. Pederson KI, Fleury BH, Mogensen PE (1997) High resolution of electromagnetic waves in time-varying channels. IEEE Proc., PIMRC'97, 4(9):650-654

8. Shiu DS, Foschini GJ, Gans MJ, Kahn JM (2000) Fading correlation and its effect on the capacity of multielement antenna systems. IEEE Trans. Commun., 48(3):502-513

9. Raleigh GG, Cioffi JM (1998) Spatio-temporal coding for wireless communications. IEEE Trans. Commun., 46(3):357-366

10. Catreux S, Driessen PF, and Greenstein LJ (2001) Attainable throughput of an interference-limited multiple-input multiple-output (MIMO) cellular system. IEEE Trans. Commun. 49(8):1307-1311

11. Chryssomallis M (2000) Smart antennas. IEEE Antennas and Propagation Magazine, 42(3):129-136

12. Winters JH (1987) On the capacity of radio communication systems with diversity in a Rayleigh fading environment. IEEE J.Select. Areas Commun., 5(6):871-878

13. Foschini GJ, Gans MJ (1998) On limits of wireless communications in a fading environment when using multiple antennas. Wireless Personal Commun., 6(3):311-335

14. Shannon CE (1949) Communication in the presence of noise. Proc. IRE., 37(1):10-21

15. Kovalyov IP (2001) Peredacha informatsyi v radiokanalakh pri ispolzovanii prostranstvennykh garmonik (Data transfer in wireless channels with use of space harmonics). Radiotekhnika, 51(5):17-20

16. Garg Vijay K (2000) IS-95 CDMA and cdma 2000, Prentice Hall, Upper Saddle River, NJ

17. Kazenelenbaum BZ, Sivoy AN (1989) Elektrodinamika antenn s poluprozrachnymi poverkhnostyami. Metody konstruktivnogo sinteza. (The electrodynamics of antennas with semi-transparent surfaces. Methods of constructive synthesis.) Nauka, Moscow

18. Korn GA, Korn TM (1968) Mathematical handbook for scientists and engineers. McGraw-Hill , New York

19. Par Andre Angot (1957) Complements de mathemaiques al'usare des ingenieurs de l'elektronique et des telecommunications. Paris

20. Karasawa Y, Iwai H (2000) Formulation of spatial correlation statistics in Nakagami-Rice fading environments. IEEE Trans. Commun., 48(1):12-18

21. Abramowitz M, Stregun IA (1964) Handbook of mathematical functions with formulas, graphs and mathematical tables. NBS

22. Vilenkin NYa (1991) Spetsyalniye funktsii i teoriya predstavleniya grupp. (Special functions and group representation theory.) Nauka, Moscow

23. Nikolsky VV (1978) Elektrodinamika i rasprostraneniye voln. (Electrodynamics and wave propagation.) Nauka, Moscow

24. Markov GT, Chaplin AF (1983) Izlucheniye elektromagnitnykh voln (Radiation of electromagnetic waves.) Radio i Svyaz, Moscow

25. Honl H, Maue AW, Westfahl K (1961) Theorie der Beugung. Springer, Berlin Heidelberg New York

Index

Printing: Druckhaus Berlin-Mitte GmbH
Binding: Buchbinderei Stein & Lehmann, Berlin